Structural Identification of Constructed Systems

Approaches, Methods, and Technologies for Effective Practice of St-Id

SPONSORED BY

Committee on Structural Identification of Constructed Systems of the
Performance of Structures Committee of the Technical Activities
Division of the Structural Engineering Institute of ASCE

EDITED BY
F. Necati Çatbaş
Tracy Kijewski-Correa
A. Emin Aktan

Published by the American Society of Civil Engineers

Library of Congress Cataloging-in-Publication Data

Structural identification of constructed systems : approaches, methods, and technologies for effective practice of St-Id / sponsored by Committee on Structural Identification of Constructed Systems of the Performance of Structures Committee of the Technical Activities Division of the Structural Engineering Institute of ASCE ; edited by F. Necati Çatbas, Tracy Kijewski-Correa, A. Emin Aktan.
   pages cm
   Includes bibliographical references and index.
   ISBN 978-0-7844-1197-1 (paper) -- ISBN 978-0-7844-7647-5 (PDF) 1. Structural analysis (Engineering)--Research. 2. Structural analysis (Engineering)--Mathematical models. 3. Building. 4. Bridges--Design and construction. 5. Building--Safety measures. 6. Bridges--Safety measures. I. Catbas, F. Necati. II. Kijewski-Correa, Tracy Lynn. III. Aktan, A. E. IV. Structural Engineering Institute. Committee on Structural Identification of Constructed Systems.
   TA646.S765 2013
   624.1--dc23
                                         2013008190

American Society of Civil Engineers
1801 Alexander Bell Drive
Reston, Virginia, 20191-4400

www.pubs.asce.org

Any statements expressed in these materials are those of the individual authors and do not necessarily represent the views of ASCE, which takes no responsibility for any statement made herein. No reference made in this publication to any specific method, product, process, or service constitutes or implies an endorsement, recommendation, or warranty thereof by ASCE. The materials are for general information only and do not represent a standard of ASCE, nor are they intended as a reference in purchase specifications, contracts, regulations, statutes, or any other legal document. ASCE makes no representation or warranty of any kind, whether express or implied, concerning the accuracy, completeness, suitability, or utility of any information, apparatus, product, or process discussed in this publication, and assumes no liability therefore. This information should not be used without first securing competent advice with respect to its suitability for any general or specific application. Anyone utilizing this information assumes all liability arising from such use, including but not limited to infringement of any patent or patents.

ASCE and American Society of Civil Engineers—Registered in U.S. Patent and Trademark Office.

Photocopies and permissions. Permission to photocopy or reproduce material from ASCE publications can be obtained by sending an e-mail to permissions@asce.org or by locating a title in ASCE's online database (http://cedb.asce.org) and using the "Permission to Reuse" link. *Bulk reprints.* Information regarding reprints of 100 or more copies is available at http://www.asce.org/reprints.

Copyright © 2013 by the American Society of Civil Engineers.
All Rights Reserved.
ISBN 978-0-7844-1197-1 (paper)
ISBN 978-0-7844-7647-5 (PDF)
Manufactured in the United States of America.

Committee on Structural Identification of Constructed Systems of the Technical Activities Division of the Structural Engineering Institute of ASCE

MEMBERS

F. Necati Catbas, Chair; formerly Chair of the Subcommittee in charge of preparing the State-of-Art Report, University of Central Florida, USA
Tracy Kijewski-Correa, Vice-Chair; formerly Officer in charge of the quality assurance of the State-of-the-Art Report, University of Notre Dame, USA

A. Emin Aktan, Founding Chair of the Committee, Drexel University, USA
Sreenivas Alampalli, New York State Department of Transportation, USA
Paul J. Barr, Utah State University, USA
Raimondo Betti, Columbia University, USA
James MW Brownjohn, Sheffield University, UK
Guido De Roeck, Katholieke Universiteit Leuven, Belgium
Gregory L. Fenves, University of Texas, Austin, USA
Dan Frangopol, Lehigh University, USA
Hitoshi Furuta, Kansai University, Japan
Kirk A. Grimmelsman, University of Arkansas, USA
Marvin W. Halling, Utah State University, USA
Ahsan Kareem, University of Notre Dame, USA
Franklin L. Moon, Drexel University, USA
Sami F. Masri, University of Southern California, USA
Babak Moaveni, Tufts University, USA
Shamim Pakzad, Lehigh University, USA
Frederick R. Rutz, J.R. Harris Co., USA
Masoud Sanayei, Tufts University, USA
Ayman A. Shama, Parsons Engineering, USA
Ian F.C. Smith, EPFL Lausanne, Switzerland
Hoon Sohn, KAIST, South Korea
Yang Wang, Georgia Institute of Technlogy, USA
Zhishen Wu, Ibaraki University, Japan

FRIENDS OF THE COMMITTEE

Jim Beck, California Institute of Technology, USA
Erin Santini-Bell, University of New Hampshire, USA
Juan Caicedo, University of South Carolina, USA
Joel P. Conte, University of California, San Diego, USA
Reginal DesRoches, Georgia Institute of Technlogy, USA
Donald Dusenberry, Simpson Gumpertz & Heger, Inc
Shirley Dyke, Purdue University, USA
Chuck Farrar, Los Alamos National Laboratories, USA
Yozo Fujino, University of Tokyo, Japan
Hamid Ghasemi, Federal Highway Administration, USA

Mustafa Gul, University of Central Florida, USA
Andrew Hermann, Hardesty-Hanover, USA
Keith Hjelmstad, Arizona State University, USA
Stefan Hurlebaus, Texas A&M University, USA
Erik Johnson, University of Southern California, USA
Hui Li, Harbin Institute of Technology, China
Yong Lu, University of Edinburgh,UK
Satish Nagarajaiah, Rice University, USA
Y.Q. Ni, The Hong Kong Polytechnic University, PRC
Qin Pan, Bechtel Engineering, USA
Jonsong Pei, University of Oklahoma, USA
Jim Rossberg, American Society of Civil Engineers, USA
Harry "Tripp" Shenton, University of Delaware, USA
Mahendra Singh, National Science Foundation, USA
Andrew Smyth, Columbia University, USA
Ertugrul Taciroglu, University of California, Los Angeles, USA
Björn Täljsten, Luleå Univ. of Technolonogy, Sweden
John Wallace, University of California, Los Angeles, USA
Helmut Wenzel, VCE Engineering Vienna, Austria
Sharon Wood, University of Texas, Austin, USA
Hae-Bum "Andrew" Yun, University of Central Florida, USA

Contents

Preface vii
Principal Contributors xiii

1   Structural Identification of Constructed Systems 1
    1.1   Overview 1
    1.2   Objectives 4
    1.3   Historical Development 4
    1.4   Six Steps for St-Id of Constructed Systems 7
    1.5   Outline of Report 12
    1.6   References 13

2   A Priori Modeling 18
    2.1   Introduction 18
    2.2   Classification of A Priori Models 18
    2.3   Common Model Types 19
    2.4   Modeling Constraints Unique to St-Id 21
    2.5   Construction of A Priori Models through 3D CAD 22
    2.6   Quality Control Requirements 23
    2.7   Closing Remarks 25
    2.8   References 25

3   Experimental Considerations 26
    3.1   Introduction 26
    3.2   Classification of Experiments Based on Input 28
    3.3   Sensors and Sensor Classification 35
    3.4   Data Transmission 47
    3.5   Data Acquisition and Management 48
    3.6   Use of Non-Destructive Evaluation for Structural Identification 51
    3.7   Closing Remarks 52
    3.8   References 52

4   Data Processing and Direct Data Interpretation 65
    4.1   Introduction 65
    4.2   Examples of Non-Physical Numerical Models 67
    4.3   Closing Remarks 72
    4.4   References 72

**5   Structural Identification for Selection, Application, and Calibration
    of Physics-Based Models** 78
    5.1   Model Selection 78
    5.2   Model Application for Structural Identification 80
    5.3   Finite Element Model Calibration 87
    5.4   Closing Remarks 95

	5.5	References 96
6	**Utilization of St-Id for Assessment and Decision Making............................ 113**	
	6.1	Introduction... 113
	6.2	Performance-Based Engineering 114
	6.3	Risk-Based Decision Making 115
	6.4	Quantitative vs. Qualitative Risk Assessment 117
	6.5	Closing Remarks... 118
	6.6	References... 118

Appendix A: Case Studies on the Structural Identification of Buildings........... 119
    A.1   Chicago Full-Scale Monitoring Program.. 119
    A.2   Four Seasons Building 126
    A.3   Three-Story Concrete Building in CSMIP... 130
    A.4   Guangzhou New TV Tower.. 140
    A.5   Seven-Story RC Building Slice 148
    A.6   Building Substructure Example: Composite Structural Floor System.. 155
    A.7   References... 163

Appendix B: Case Studies on the Structural Identification of Bridges 169
    B.1   Henry Hudson Bridge 169
    B.2   Throgs Neck Bridge 174
    B.3   Golden Gate Bridge 181
    B.4   Vincent Thomas Bridge 188
    B.5   Hakucho Suspension Bridge 193
    B.6   Yokohama Bay Bridge.. 196
    B.7   Alfred Zampa Memorial Bridge 201
    B.8   Langensand Bridge 205
    B.9   Sunrise Boulevard Movable Bridge... 209
    B.10   New Svinesund Bridge 217
    B.11   References... 222

Index 227

Preface

ASCE-SEI Performance of Structures Track established the current technical committee on "Structural Identification (St-Id) of Constructed Systems" in 2005. This was intended to continue with the unfinished work of an earlier ASCE Committee with the same name that was established over a decade ago but discontinued after 1999. The previous Committee issued a draft report that was circulated but not published, and many other contributions were made by the Committee in terms of bringing researchers from engineering mechanics and civil-structural engineering closer in the understanding of the powerful paradigm of St-Id especially in the context of applications to constructed systems. The contributors to the previous report included distinguished researchers such as James Yao, James Beck, Scott Doebling and Chuck Farrar amongst others, who are well-known as pioneers who have explored and contributed to St-Id of constructed systems, albeit mainly from an engineering mechanics world-view.

The current Committee, founded by the founding Chair of the earlier one, recognized that it is critical to understand and leverage the concept of St-Id mainly from a structural engineering world-view in order to advance the art of structural engineering. Differences between engineering mechanics versus structural engineering world-views is not often discussed, however, a review of the Table of Contents of the respective Journals of these sub-disciplines of civil engineering reveals a gap that is difficult to deny. One of the goals of the 2005 Committee was to try to bridge this gap. Given the above, the near-term purpose and objectives of the Committee were defined as follows:

Committee Purpose: Foster advances and dialogue to enable the collection of data, its analysis and interpretation, and ultimately the assessment of constructed system performance beyond the anecdotal observations that currently form our bases for judging the merits of our designs.

Committee Objectives: Given that the actual mechanical characteristics and performance of constructed systems have been shown to be very different from those considered during a specification-based code design, a principal focus will be on defining metrics and establishing measurement standards for constructed systems, which represents a prerequisite for a meaningful transition to Performance Based Civil Engineering. The Committee will also aim to develop guidelines for reliable field-calibrated analytical modeling and characterization of existing constructed systems.

Field-calibrated analytical modeling leverages objective measurements of geometry, soil and structural material characteristics, responses during controlled experiments, and, long-term monitoring for establishing the loading environments and performance of a constructed system at critical performance limit-states. To inform these two standards as well as the profession at large regarding the relevant issues and vast discrepancies between different applications of St-Id, the Committee will collect available data from existing tests, and interpret and archive case studies.

In addition to preparing and publishing guidelines, conference sessions will be organized and papers will be published in proceedings.

The following State-of-the-Art report on Structural Identification is the Committee's first major product for the profession to fulfill the promise of both the earlier and the current Committee.

Background and the Drivers Shaping the Agenda of the Committee

Structural identification (St-Id) is an adaptation of the system-identification concept which originated in electrical engineering in relation to circuit and control theory. St-Id has been defined as: *"the parametric correlation of structural response characteristics predicted by a mathematical model with analogous quantities derived from experimental measurements"* (the 1995 draft report by the earlier Committee). The St-Id paradigm was first introduced to engineering mechanics researchers by Hart and Yao (1977) and to civil-structural engineering researchers by Liu and Yao (1978). These seminal papers gradually inspired many researchers to investigate various aspects of St-Id, and nearly 30 years later St-Id remains an active research area in both engineering mechanics and civil-structural engineering. Recent advances in IT has rendered FE modeling of large structures for new design, or condition and vulnerability assessment, rehabilitation or retrofit commonplace. Civil engineering consultants are routinely using FE modeling and simulation for practical applications. However, it has been well established that due to the uniqueness and significant epistemic uncertainty associated with our constructed systems, reliable simulations, either by a 3D microscopic FE model or by much simpler and greatly idealized macroscopic models, require calibration and validation based on actual observations and measured experimental data. Meanwhile, the paradigm of making meaningful observations and taking reliable measurements from actual operating constructed systems in the field is still an emerging art. Using field observations and measurements for calibrating and validating a FE model is also a highly challenging problem. An ASCE Committee was therefore warranted to bring together researchers and practicing engineers with experience and knowledge in practical applications of FE models, fundamentals of various approaches to FE modeling, simulations and scenario analysis, field research, and their integration

St-Id Application Scenarios

There are several scenarios, which may justify the construction and identification, based on the results of field experiments, of an analytical model for simulating an actual constructed system. Examples include:

1. Design verification and construction planning in case of challenging and/or ground-breaking new designs,
2. A means of measurement-based delivery of a design-build contract in a performance-based approach,
3. Documentation of as-is structural characteristics to serve as a baseline for assessing any future changes due to aging and deterioration, following hazards, etc.
4. Load-capacity rating for inventory, operations or special permits,

5. Evaluation of possible causes and mitigation of deterioration, damage and/or other types of performance deficiencies (e.g. vibrations, cracking, settlement, etc.),
6. Structural intervention, modification, retrofit or hardening due to changes in use-modes, codes, aging, and/or for increasing system-reliability to more desirable levels,
7. Health and performance monitoring for operational and maintenance management of large systems,
8. Asset management of a population of constructed systems such as RC T-Beam bridges,
9. Advancing our knowledge regarding how actual structural systems are loaded (during construction and after commissioning), how they deform, i.e. their kinematics at supports, joints, connections, and how they transfer their forces through the members to foundations and to soil.

There is sufficient evidence that our current knowledge base on the loading, behavior and performance of constructed systems is greatly incomplete, especially when new construction materials and systems are considered. The significant epistemic uncertainty prevailing in the actual loading mechanisms, intrinsic force distributions, kinematics, failure modes and capacities of existing constructed systems, especially after aging may lead to discrepancies between predicted responses and capacities that are different by more than an order of magnitude, and not always in a conservative way. Many members, joints and connections may be loaded less than assumed while many others may be loaded with demands that are far greater than anticipated. Some of the mechanisms that may control the distribution of demands and the corresponding capacity at the critical regions of constructed systems are often very difficult to discover and quantify even with measurements unless a rigorous St-Id is carried out by experts. Therefore, systematic applications of St-Id to well selected samples of constructed systems is considered important if civil engineering in the USA will move from specification-based project delivery to a performance-based one with long-term warranty.

Observations in the field, followed by properly designed, executed and interpreted experiments are the only definitive approach for reducing epistemic uncertainty that clouds constructed system behavior. Further, it is not possible to reliably design, execute and interpret field experiments without first studying, observing, conceptualizing, and modeling a constructed system, so that sensing and loading can be designed effectively and data can be interpreted. These are some of the reasons that make advancing St-Id and conducting applications important and in fact necessary for civil engineers if we are to respond to the needs of the society regarding improving the lifecycle performance and sustainability of our constructed systems. The Committee has developed the following long-term work-plan to accomplish its goals:

1. **Recruiting Champion Experts:** The first step has been to bring together as many of the champions and experts as possible from academe, industry and government who have been advocates in changing the way we teach and practice civil engineering. Such experts would be the first to recognize the

importance of St-Id as a paradigm offering an effective path to integration and discovering the reality of constructed systems and infrastructures. Civil engineers knew of this reality before the 20th Century through intuition and heuristics. After losing this to a proliferation of university programs and prescriptive codes; and to the shift to applied science in the 1950's without distinguishing the differences between constructed and mechanical systems, we now have a chance to rediscover reality. The most important long-term goal for the Committee would be to establish reality of civil engineered systems in a factual and quantitative manner, as accurately and completely as possible, comparing the assumed-predicted and true reality, and to disseminate this information. In this manner we may contribute to changing the way we teach and practice by basing it on ground truth which we can discover through St-Id of existing systems. The Committee membership has been endowed with the best possible global expertise.

2. **Establishing the State-of-the-Art:** The first deliverable identified by the Committee is a state-of-the-art report that is now complete and that recognizes the distinctions between constructed and mechanical systems.

3. **Organizing/Coordinating Research on Lab Benchmarks:** Well-designed physical laboratory models are invaluable benchmarks for exploring and demonstrating the St-Id process, and the many possible products that may come out of the process. Such models also serve as excellent case-study based learning opportunities for students and practicing engineers interested in continuing education. Recognizing that there is no unique characterization for a constructed system, but a ground truth that we can approach only as close as uncertainty permits, the art of St-Id becomes how we deal with the challenges of managing uncertainty. **The age-old strategy in civil engineering analysis has been investing only as much into modeling and computation that is commensurate with what the uncertainty will permit us to predict within some confidence.** The issue is in how we may reduce the uncertainty by virtue of having a physical model that does NOT have many of the uncertainties we face in the field, and one that may be tested as many times, by as many persons, and, in as many different ways as needed.

4. **Organizing/Coordinating Demonstrations on Real Systems:** A long-term goal is to leverage the expertise of the Committee in St-Id demonstration projects on real bridges and buildings. Such an effort may be initiated in NIST's leadership for buildings and FHWA's leadership for bridges, with ASCE and other agencies such as NSF's participation and support.

Acknowledgements

The Committee acknowledges a prior ASCE-SEI Committee (1997-2002) in which the Chair of the current Committee served under the pioneer who introduced the concept of St-Id to the civil engineering community, the late Dr. J.T.P. Yao. Writers deeply acknowledge the leadership of Dr. J.T.P. Yao, and the contributions of the members of the former ASCE Committee on St-Id of Constructed Facilities.

Special thanks are overdue to Dr. Doebling and Dr. Helmicki who have made major contributions to the draft report of the earlier Committee.

    Dr. A. E. Aktan, *Chair of the Committee*
    Dr. Necati Catbas, *Chair of Subcommittee in charge of preparing the State-of-Art Report*
    Dr. Tracy Kijewski, *Officer in charge of the quality assurance of the State-of-the-Art Report*
    Dr. Franklin Moon, *Secretary of the Committee*

Principal Contributors

Integration and Editing:

Necati Catbas, *University of Central Florida*, and **Tracy Kijewski-Correa**, *University of Notre Dame*

Chapter Leads & Contributors

Chapter 1: **Franklin Moon**, *Drexel University* and **Necati Catbas**, *University of Central Florida*

Chapter 2: **Franklin Moon**, *Drexel University* and **Necati Catbas**, *University of Central Florida*

Chapter 3: **James Brownjohn**, *University of Sheffield*, **Kirk Grimmelsman**, *University of Arkansas*, **Hoon Sohn**, *KAIST/Carnegie Mellon University*

Chapter 4: **Ian Smith**, EPFL, *Swiss Federal Institute of Technology, Switzerland*, **Chuck Farrar**, *Los Alamos National Laboratories*, **Hoon Sohn**, *KAIST/Carnegie Mellon University*

Chapter 5: **Ian Smith**, EPFL, *Swiss Federal Institute of Technology, Switzerland*, **Erin Bell**, *University of New Hampshire*, **Masoud Sanayei**, *Tufts University,* **Chuck Farrar**, *Los Alamos National Laboratories*

Chapter 6: **Franklin Moon**, *Drexel University* and **Dan Frangopol**, *Lehigh University*

Appendix A: **Tracy Kijewski-Correa**, *University of Notre Dame*

Appendix B: **Necati Catbas**, *University of Central Florida*

Case Study Contributors

Chicago Full-Scale Monitoring Program: **Tracy Kijewski-Correa**, *University of Notre Dame*

Four Seasons Building: **Roshanak Omrani**, **Ertugrul Taciroglu**, *University of California at Los Angeles*

Three-Story Concrete Building in CSMIP: **Roshanak Omrani**, **Ertugrul Taciroglu**, *University of California at Los Angeles*

Guangzhou New TV Tower: **Y.Q. Ni**, *Hong Kong Polytech University*

Seven-Story RC Building Slice: **Babak Moaveni**, *Tufts University,* **Xianfei He**, *AECOM Transportation*, **Joel Conte**, *University of California, San Diego*

Building Substructure Example-Composite Structural Floor System: **James MW Brownjohn, Paul Reynolds** and **Alex Pavic**, *Sheffield University, UK*

Henry Hudson Bridge: **Jian Zhang, Qin Pan** and **Franklin Moon**, *Drexel University*

Throgs Neck Bridge: **John Prader, Jian Zhang** and **Franklin Moon**, *Drexel University*

Golden Gate Bridge: **Shamim Pakzad**, *Lehigh University*, **George Fenves**, *University of Texas, Austin*

Vincent Thomas Bridge: **Hae-Bum Yun**, *University of Central Florida*, **Sami Masri**, *University of Southern California*

Hakucho Suspension Bridge: **Yozo Fujino, Dionysius Siringoringo, Tomonori Nagayama**, *University of Tokyo*

Yokohama Bay Bridge: **Yozo Fujino, Dionysius Siringoringo, Tomonori Nagayama**, *University of Tokyo*

Alfred Zampa Memorial Bridge: **Xianfei He**, *AECOM Transportation*, **Babak Moaveni**, *Tufts University*, **Joel Conte**, *University of California, San Diego*

Langensand Bridge : **James-A. Goulet, Ian Smith**, EPFL, *Swiss Federal Institute of Technology, Switzerland*

Sunrise Boulevard Movable Bridge: **Necati Catbas, Mustafa Gul**, *University of Central Florida*

Svinesund Bridge: **Hendrik Schlune, Mario Plos, Kent Gylltoft**, *Chalmers University of Technology, Sweden*

Acknowledgements for Report Editing Assistance

Necati Catbas (University of Central Florida) would like to acknowledge his graduate student Mustafa Gul of UCF for his assistance in organizing multiple files for different chapters, editing as well as for his input on technical issues. Tracy Kijewski-Correa (Notre Dame) wishes to acknowledge the assistance of undergraduate students Michael Wodarcyk and Antonio Ayala and graduate students Jennifer Cycon and Jeffrey Loftus of Notre Dame for their assistance in compiling and databasing references for this report.

Chapter 1
Structural Identification of Constructed Systems

1.1 Overview

For centuries the civil/structural engineering profession (under various names) was forced to rely on extremely simplistic and idealistic models of constructed systems for analysis and design. Buildings were modeled as plane frames and bridges as simple or continuous beams. This simplistic view of constructed systems forms the basis of many of our process-based and prescriptive codes, and although much more sophisticated modeling approaches are readily available, many engineers continue to opt for simplicity. These approaches, when coupled with current codes, and especially when applied by engineers with experience and following sound heuristic principles, have proven quite capable of developing economic, safe designs. The primary shortcoming of these approaches is their inability to accurately simulate the actual performance of constructed systems. As the profession moves towards more performance-based design approaches, and begins to more seriously consider durability, maintenance, and serviceability limit states as well as struggle with contemporary challenges associated with preservation and renewal, such simple modeling approaches are inadequate.

Today there are powerful modeling tools available that have the ability to simulate both the three-dimensional local and global behaviors of constructed systems. Unfortunately, the challenge of reliably simulating the performance of constructed systems requires far more than the adoption of more refined models. There are several examples that show even very detailed models may miss critical mechanisms and force distributions within a complex structure, and indicate that discrepancies between measured and simulated responses can be as high as 100% and 500% for global and local responses, respectively. The reality is that although such refined models have the ability to simulate behavior with more resolution, they require far more information to mitigate the influences of bias (epistemic, i.e. due to lack of sufficient data and approximate models) sources of uncertainty (e.g., boundary and continuity conditions) for reliable results. This understanding has resulted in a growing recognition for the need to improve model predictions using experimental response data (considering the confluence of experimental, analytical, and numerical/computational errors), and has fueled the on-going development of Structural Identification (St-Id).

The paradigm of St-Id aims to bridge the gap between the model and the real system by developing reliable estimates of the performance and vulnerability of structural systems through the improve simulations using experimental observations/data. St-Id is the process of creating/updating a physics-based model of a structure (e.g., finite element model) based on its measured static and/or dynamic measured response which will be used for assessment of structure's health and performance and as well as decision making. St-Id is a transformation of system identification which focuses on creating a numerical or non-physics-based model (e.g., difference equation, state-space) of a dynamic system based on its measured response. In the case of manufactured systems (airplanes, cars and space structures, etc.) St-Id has become common practice and has proven to be a dependable tool for understanding actual mechanical characteristics and for informing both design and performance assessment. In contrast, for constructed systems (building, bridges, dams, etc.) St-Id remains in its infancy and has enjoyed only sparse implementation in practice. In the hopes of increasing its appropriate use, the primary objective of this report is to provide the engineering community with a broad overview of recent advances that are enabling reliable applications of St-Id to constructed systems. While many challenges remain, it is becoming increasingly clear that St-Id has the potential to pay enormous dividends by providing a direct link between the engineering profession and the constructed systems they design, construct, operate and preserve.

As demonstrated time and again, the uniqueness and uncertainties associated with constructed systems render their actual mechanical characteristics and performance parameters extremely difficult to predict in an a priori sense and even simulate following an experiment. Along with challenging St-Id, this clearly demonstrates how little is truly known about the performance of actual constructed systems. Consider that current design and assessment procedures do not have any quantitative linkage to actual constructed systems; such gaps are filled by qualitative observations, anecdotal experiences, or laboratory studies rooted in reductionism. So while the challenges are substantial, the paradigm of St-Id holds great promise to uncover the reality of constructed systems. This uncovering has broad implications that range from informing and underpinning performance-based design approaches to facilitating quantitative risk-based decision-making for aging infrastructure systems (Aktan and et al. 2007). By offering a rational means to collect, analyze, and interpret quantitative data from constructed systems, St-Id has the potential to reduce the need for excessive conservatism in the face of uncertainty, and to expand the assessment of structural performance beyond its current, exclusive reliance on visual appearance.

While some have argued that the current lack of widespread St-Id applications to constructed facilities is primarily due to a lack of practical sensing technology, recent advances in this area have not yet been accompanied by a significant increase in applications. Rather, the authors believe the barrier is more fundamental in nature. Over the last decade, many have unsuccessfully attempted to apply St-Id approaches developed and proven in the case of manufactured systems directly to constructed systems. In many cases these applications resulted in incremental, if any, benefits to owners and have fueled skepticism regarding the usefulness of St-Id. The authors believe that the underlying challenge is that St-Id approaches developed for manufactured systems implicitly ignore the many unique and confounding attributes of constructed systems (Table 1-1).

Recognizing this, some of the most successful applications of St-Id to constructed systems have developed modifications that explicitly address some of these attributes. For example, the recent progress associated with operational modal analysis explicitly recognizes the cost and difficulty of performing forced vibration tests on large constructed systems (Brownjohn et al. 1992; Fujino et al. 1999; Wenzel

STRUCTURAL IDENTIFICATION OF CONSTRUCTED SYSTEMS 3

and Pichler 2005). In addition, several researchers have developed St-Id approaches that explicitly address aleatory uncertainty, which can be significant for constructed systems (Beck 1990; Beck and Katafygiotis 1998; Bucher et al. 2003; Yuen and Katafygiotis 2002). While these advances are important, the distinctions between constructed and manufactured systems they address are far from exhaustive, and thus a wider, sustained effort is necessary. Such an effort must be multi-disciplinary in nature and must begin with a clear articulation of the state-of-the-art related for the St-Id of constructed systems.

Table 1-1. Uncertainties unique to constructed systems that influence their mechanical characteristics and performance

Heterogeneity	Materials, member proportions, detailing, etc can vary considerably from member to member, and within a member. Deterioration and damage compounds these variations and makes discretization difficult and sometimes unmanageable without heuristics.
Boundaries	Constructed systems have unobservable soil-foundation interfaces that are often non-stationary in their contact properties. Soil and even rock properties change with pressure, moisture, temperature and time.
Continuity	Most constructed systems, and especially bridge systems are designed with movement systems and/or force releases. These systems are most often unobservable and behave differently under different levels of force and temperature.
Redundancy	Constructed systems have many types of local, regional and global/external redundancies. These redundancies are highly affected by temperature changes and temperature gradients (due to radiation), which results in intrinsic forces and changes in element properties.
Intrinsic Forces	Constructed systems maintain complex and non-stationary intrinsic forces due to dead weight, construction loads/staging, temperature effects, deterioration, damage, overloads, etc. These intrinsic forces are nearly impossible to measure in an absolute sense and their changes in many cases overwhelm the forces due to transient live loads.
Types of Nonlinearity	Element, connection and global behavior of real constructed systems exhibit many different types of nonlinearity that change at different limit-states. Cracking, material yielding, local instability, connection slip, interface friction, etc. are all associated with hardening/softening type behaviors.
Non-Stationarity	Constructed systems are non-stationary due to the non-stationary nature of their environment (e.g., temperature, radiation, etc) as well as their various loading-level and loading-type related nonlinearities. Temperature and humidity effects are highly complex: changes and rate of changes in ambient, regional and local temperatures and humidity of the structure and the soil may lead to intrinsic forces and also induce changes in boundary and continuity conditions.
Uniqueness	Nearly all constructed systems are custom-designed for specific applications and their mechanical characteristics are strongly affected by events during and immediately following their construction. While types of constructed systems may be grouped based on their primary structural system, size, materials, etc., applying results from a single structure to a larger population of structures is challenging due to their inherent uniqueness.
Geometric, Temporal Scale, Cost, Lifecycle	Constructed systems such as major highway bridges or combinations of several bridges and tunnels within regional transportation networks may be longer than several miles, cost several billions of dollars, and be expected to remain in service for over 100 years. The size and lifecycle impedes our ability to view such systems in a holistic manner over a sufficient span along their lifecycles and further compounds the natural variability and uncertainty in their mechanical characteristics.
Coupling	Most constructed systems feature coupling between sub-systems such as frames and walls, water, soil and foundation, substructure and superstructure, structural and nonstructural, etc. The coupling between sub-systems maybe highly complex, often nonlinear and nonstationary. Yet coupling may control how forces and displacements are transferred and the responses.

1.2 Objectives

Given the benefits, as well as the significant challenges alluded to above, the St-Id of constructed systems has attracted the attention of numerous researchers worldwide over the last several decades. It is the goal of this report to benchmark and provide an overview of these developments, which constitute the current state-of-the-art. A primary contribution of any such effort is in structuring the field and providing categories, which will serve to delineate and locate different developments and lines of inquiry in context. While this, for the sake of completeness, will require brief descriptions of on-going and unfinished research, the aim of this report is to focus on established methodologies and approaches that have matured to the point where they may be employed in practice.

This report is written mainly for practicing engineers and is intended to provide an objective view of what sound applications of St-Id can and cannot offer. In this manner, this report aims to provide practicing engineers with the ability to decide whether St-Id may be appropriate in certain situations and to guide them as to what may be expected from such applications. It is emphasized that this report is not a 'how to' manual or a 'best practices' document, and was not assembled with the details or the intent to support actual applications.

1.3 Historical Development

The origins of St-Id are soundly founded on the scientific method of observation (experiment), hypothesis (model) and validation. This approach is rooted in the argument that the establishment of reality requires that mathematical models be hypothesized and validated based on observations of the physical world. According to (Eves 1990), this concept dates back to Plato (380 B.C.) who claimed that "the reality which scientific thought is seeking must be expressible in mathematical terms, mathematics being the most precise and definite kind of thinking of which we are capable". Up to modern times, there is evidence that master builders took advantage of this concept by observing the performance of other and their structures and in some cases even conducting experiments, which no doubt influenced their methods of calculation. In 1907 Robert Maillart argued that designers should be encouraged to check their analytical assumptions through load tests that establish deflections of the completed structure (Billington 1979). Load testing of bridges has been questioned in many countries due to the difficulty of interpreting measurement data. Simple model calibration techniques that do not account for uncertainty and parameter compensation reveal unrealistic values for parameters. Rather than try to explain discrepancies, engineers have decided to stop load testing. It is a shame that over a century later such rational and reasonable advice has not been heeded in a comprehensive manner.

The contributions of such pioneering efforts notwithstanding, it is important to draw a distinction with modern applications of St-Id that have their origins in systems engineering, which began in earnest during the late 1950's. During this time extreme pressure was placed on the military and their civilian contractor teams to develop, test, and place in operation nuclear tipped missiles and orbiting satellites. At the same time, systems engineering was also evolving in the commercial telecommunications sector. The advent of the computer permitted extensive simulation and evaluation of systems, subsystems, and components; thus accurate synthesis of system elements became possible. These advances contributed to the development of the system identification (Sys-Id) concept, defined as the estimation of a system based on the correlation of inputs and outputs, which originated in electrical engineering in relation to circuit and control theory. Paralleling the technological boom of the past several decades, the Sys-Id concept has flourished as numerous engineering disciplines have recognized its value. Today, Sys-Id serves as a fundamental prerequisite for addressing systems problems in mathematics, physics, economics, social sciences, and throughout engineering (Kossiakoff and Sweet 2002).

Structural identification (St-Id) is a transformation and application of Sys-Id to mechanical (manufactured) and civil (constructed) structural systems. The paradigm was first introduced to engineering mechanics researchers by Hart and Yao (1977) and to civil-structural engineering researchers by Liu and Yao (1978). These seminal papers gradually inspired many researchers to investigate various aspects of St-Id, and over 30 years later St-Id remains an active research area in both engineering mechanics and civil-structural engineering. The following sections provide a cursory description of St-Id research and developments over the last several decades. This discussion is structured based on the type of model employed (Table 1-2). For the interested reader, far more comprehensive historical information related to various aspects of St-Id research have been documented in numerous literature survey papers and reports (Doebling et al. 1998; Hudson 1970; Hudson 1976; Ibanez 1979; Moon and Aktan 2006; Mottershead and Friswell 1993; Sohn et al. 2004).

Table 1-2. Classification of analytical modeling forms

Physics-Based Models	Non-Physics-Based Models

Mathematical Physics Models
 • F=ma
 Continua Models
 • Theory of Elasticity
 • Field and Wave Eqns
 • Idealized Diff. Eqns (Bernoulli, Vlasov, etc.)
 	
 Discrete Geometric Models
 • Smeared-Macro or Element Level Models
 • FEM-for Solids and Field Problems (*most commonly used by practicing engineers*)
 • Modal Models:
   - Modal Parameters
   - Ritz Vectors
 Numerical Models
 • K,M,C Coefficients
	Semantic Models
	 • Ontologies
	 • Semiotic Models
	 Meta Models
	 • Input-Output Models
	 • Rule-based Meta Models
	 • Mathematical (Ramberg-Osgood, etc.)
	 Numerical Models
	 • Probabilistic Models
	   - Histograms to Frequency Distribution
	   - Standard Prob. Distributions
	   - Independent events
	   - Event-based
	   - Time-based
	   - Symptom-based
	 • Agents
	 • Statistical (Data-Based)
	   - ARMA, ANN, others
	   - Signal/Pattern Analysis, Wavelet, EMD, others

1.3.1 Physics-Based Models

Since the 1970's numerous researchers have investigated the use of physics-based (PB) St-Id approaches to identify actual constructed systems (Agbabian et al. 1990; Aktan et al. 1997; Aktan et al. 1998; Biswas et al. 1989; Brownjohn et al. 1987; Brownjohn 2003; Brownjohn et al. 2003; Doebling et al. 1998; Douglas and Reid 1982; Farrar and Doebling 1998; Fujino and Abe 2002; Hornbuckle et al. 1973; Kou and DeWolf 1997; Maeck and De Roeck 2003; Natke and Yao 1986; Natke and Yao 1989; Stubbs et al. 1992; Teughels and De Roeck 2004). These models are formulated to explicitly address the boundary and continuity conditions, equilibrium and kinematics (with varying degrees of resolution depending on the model type selected) of the constructed system of interest. The primary benefit of PB approaches is that the identified model facilitates the use of heuristics and can be used to explicitly simulate behavior under various critical loading conditions. Because of this, such models can diagnose the causes of changes in behavior as well as identify how such changes may impact the performance of the overall system. While several researchers have investigated the use of nonlinear models (Chassiakos et al. 1995; Jayakumar and Beck 1988; Kapania and Park 1997; Naghavi and Aktan 2003; Smyth et al. 1999), currently the most commonly employed PB St-Id approach relies on linear matrix structural analysis or finite element (FE) models, in a more general form including matrix structural analysis.

1.3.2 Non-Physics-Based Models

Since the early 1990s researchers have been investigating the use of many different types of non-physics-based (NPB) models for St-Id, including Artificial Neural Networks (ANNs) (Chang et al. 2001; Masri et al. 1996; Nakamura et al. 1998; Zapico et al. 2003), wavelet decomposition (Al-Khalidy et al. 1997; Gurley and Kareem 1999; Hou et al. 2000; Kijewski and Kareem 2003), auto-regressive moving average vector (ARMAV) models (Andersen and Kirkegaard 1998; Bodeux and Golinval 2001; Shinozuka and Ghanem 1995), state space models, and Empirical Mode Decomposition (EMD) in conjunction with the Hilbert-Huang Transform (Huang and et al. 1998; Vincent et al. 1999; Yang et al. 2001). The main advantage of these techniques is that they are data-driven, i.e., the construction of NPB models is solely dependent on the data provided. This data driven nature makes them attractive for modeling complex phenomena, automation, real-time St-Id, continuous monitoring, and minimizing errors due to user interaction. While these benefits cannot be ignored, it is equally important to recognize that they can only identify whether a change in behavior that corresponds to the data recording process has occurred and cannot (in the absence of PB techniques) identify the cause of the change of its affect on overall performance. More importantly, until many decades of data with sufficient density and bandwidth is captured and analyzed, it will not be possible to definitively identify and differentiate between "normal" and "abnormal". Mitigation of measurement errors remains a significant and often unrecognized problem. These approaches are discussed in greater detail related to direct data interpretation (Chapter 4).

STRUCTURAL IDENTIFICATION OF CONSTRUCTED SYSTEMS   7

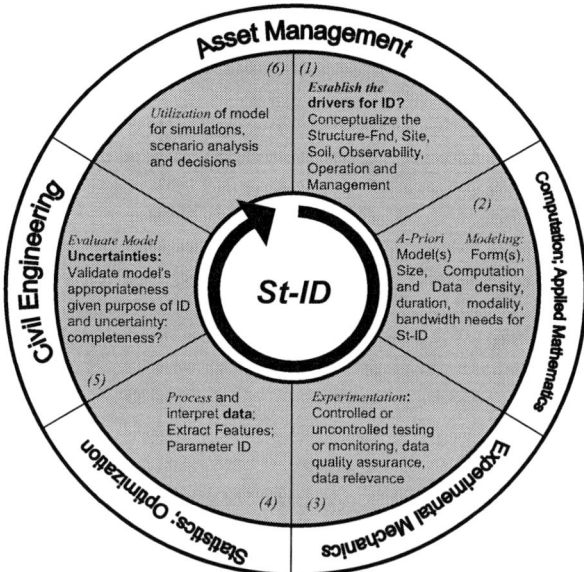

Figure 1-1. Structural identification stages

1.3.3 Combined Approaches

Over the past decade, there has been a growing awareness of the importance of incorporating uncertainty within the St-Id process for constructed systems. In response, several researchers have developed methods that explicitly incorporate the uncertainties associated with the identified modal parameters within the model updating process using PB models (Beck 1990; Beck and Katafygiotis 1998; Yuen and Katafygiotis 2002). In addition, several tools have been developed for St-Id using PB models with special emphasis on uncertainty analysis and quantification. For example, tools such as Southwest Research Corporation's NESSUS and Sandia National Laboratories' DAKOTA can be used in conjunction with commercially available FE packages to quantify uncertainty or perform sensitivity analyses, among other capabilities. While these developments are quite relevant, it should be emphasized that they exclusively address aleatory (random) uncertainty and do not provide insight into the effects of epistemic (bias) uncertainty. Currently heuristics is the only tool available to identify and mitigate the potential impacts of epistemic uncertainty on St-Id.

1.4 Six Steps for St-Id of Constructed Systems

To organize the diverse paradigm of St-Id, the ASCE St-Id of Constructed Systems Committee adopted the six steps shown in Figure 1-1 (Aktan and Moon 2005). As evidenced by these diverse steps, a team of multi-disciplinary experts is often required to fully achieve the potential of any application related to decision-support

8 STRUCTURAL IDENTIFICATION OF CONSTRUCTED SYSTEMS

or vulnerability and risk assessment. The application must be driven by experts with domain knowledge related to constructed systems and their asset management (applied systems analysis, heuristics), and should be built around a set of carefully designed objectives that are both attainable and will be of direct and demonstrable benefit. Although this is a necessary condition, successful applications of St-Id to constructed systems require additional expertise related to: modeling (analytical, numerical); experimentation (observations, sensing, data acquisition); data processing (error screening, feature extraction, etc.); comparison of model and experiment (model selection, parameter identification.); and decision-support (parametric studies, scenario analyses, risk assessment, etc.).

Today, it is not possible for any one individual to claim expertise throughout the entire spectrum of critical knowledge pertinent to the St-Id of constructed systems. It is the hope of the authors that the comprehensive framework suggested in this report will enhance appreciation for each aspect of the process, and serve as a means to facilitate more coordinated multi-disciplinary applications of St-Id. As an introduction to each step of the process, the following subsections provide brief overviews.

1.4.1 Step 1: Objectives, Observation and Conceptualization

The first step of St-Id involves becoming familiar with the issue that is driving the application as well as the structure itself. Based on the authors' experience, there are several scenarios involving the construction, operation, maintenance or lifecycle asset management that may lead owners/stewards to pursue a St-Id application. The following is a partial list.

1. Load-capacity such as for different occupational conditions, wind and earthquake loadings for buildings or rating for inventory, operations or special permits for bridges
2. Design verification and construction quality control especially in case of challenging and/or ground-breaking new designs
3. A measurement-based delivery of a design-build contract in a performance-based framework
4. Documenting the as-is structural characteristics in order to serve as a baseline for assessing any future changes, due to aging and deterioration, following hazards, etc.
5. Evaluation of possible causes and development of mitigation strategies for deterioration, damage and/or other types of performance deficiencies (e.g., vibrations, cracking, settlement, etc.)
6. Designing structural modification, retrofit or hardening due to changes in use-modes, codes, aging, and/or for increasing system-reliability to more desirable levels

While each one of the specific issues above may appear distinct, in fact the underlying driver in each case is the need to understand how constructed systems are actually loaded (during construction and after commissioning), how they deform, (i.e., their kinematics at supports, joints, connections), and how they transfer forces through their members to their foundation/soil systems. There is sufficient evidence that our current knowledge base on the loading, behavior and performance of constructed systems is greatly incomplete, especially when new construction methods, materials and structural systems are considered. For example, the uncertainty associated with the actual loading mechanisms, kinematics, and intrinsic force distributions of existing constructed systems (especially after aging) has led to predicted responses that are 10-20 times different from what is actually measured, and not always in a conservative way.

It follows that in order to properly guide a St-Id application it is critical that potential uncertainties be identified at the outset in Step 1. If the structure is not properly conceptualized in its current state it is likely that some potentially significant behavior mechanisms that have uncertainty associated with them may be under appreciated. This may lead to poor model construction (i.e., inadvertently simplifying or idealizing critical mechanisms improperly) and/or incomplete experimental design, which will in turn influence each step of the process. In such cases, the St-Id will at best result in inconclusive results and at worst in a significant lack of conservatism. It is the experience of the authors that most unsuccessful applications of St-Id fail to fully appreciate the true complexity of the constructed system in Step 1, and thus before any model is developed or any data acquired, their value is compromised.

The key to successful applications of St-Id is in the art of conceptualizing large constructed systems given that many critical details, mechanisms, episodes and behaviors are not feasible to directly observe and many of these may be impossible to reliably infer from indirect measurements. Many attributes that are necessary for reliable modeling may have been undocumented. Without complete support and full collaboration from a facility's owners and managers, and a full understanding of the operational demands and related constraints, it is not possible to expect a meaningful application of St-Id. The data, information and knowledge that are available about a system that will be identified would serve as important constraints and drivers for the analytical modeling, measurements and controlled experiments, and model-calibration. It follows that establishing the need for a particular St-Id application, the available data, information and knowledge about the facility and the constraints on the observability of the system will dictate the scope and the return on investment that may be expected from any St-Id application. There are examples of costly investment into detailed modeling and extensive experiments without a careful consideration of the above issues that comprise Step 1.

1.4.2 Step 2: Measurement, Visualization and A Priori Modeling

Measurement of geometry and 3D visualization are critical steps that are often omitted in St-Id. There is sufficient evidence that trying to use 2D plans (even if as constructed drawings are available) for actual dimensions, mass and 3D geometry of an as-is constructed system may lead to significant errors. Depending on the objectives of St-Id, investment into close-range photogrammetry transformed into 3D CAD, and 3D imaging by laser-scanning is advised. In any case, an error-mitigation strategy for any errors in existing documentation and for the inevitable human errors in building an a-priori model from existing documentation is necessary before a-priori modeling.

10 STRUCTURAL IDENTIFICATION OF CONSTRUCTED SYSTEMS

The development of an *a priori* model within the structural identification process serves to provide estimates of structural responses that will aid in the selection of appropriate experimental approaches and applications (Step 3). The actual modeling approach adopted is dependent on the objectives of the St-Id as well as the complexity of the structure being identified. In some cases idealized mechanical mass-spring models may be sufficient, whereas in other cases high resolution geometric-replica FE models may be justified. In all cases established modeling practices should be followed, e.g., the effect of modeling assumptions should be examined through the comparison of several modeling approaches and through sensitivity analyses. Depending on the objectives of the St-Id, the a priori model may also serve as the model calibrated through parameter identification (Step 5). In these cases, the overall objectives of the study may require more refined *a priori* models than could be justified to support the experimental program. Chapter 2 of this report provides a more complete discussion of Step 2.

1.4.3 Step 3: Controlled Experimentation

Experimental methods and technologies, applied within Step 3 of the St-Id process, serve as the only objective, quantitative link to the constructed system of interest. As such, this step is indispensable. However, it is important to recognize that experimentation on constructed systems is still an emerging field of research by itself. Loading/excitation, instrumentation, data acquisition, data quality assurance including pre-processing, data communication/archival, and most importantly, documentation of the overall experiment require highly specialized and multi-disciplinary training and extensive heuristics. In the past, the only linkage to actual constructed systems involved visual inspection, testing concrete cores and steel coupons retrieved from a structure, or, under unusual circumstances, strain-gauging. However, at the present time civil engineers have many dozens of options for measuring strain, linear distortion, normalized or relative deformation, tilt, velocity and acceleration among many other measurands including mechanical-thermal and electro-chemical phenomena at the microscopic scales. Many additional options for nondestructive testing and evaluation (NDE) have become available which permit imaging and identification of discontinuities and faults within members and connections.

While such major leaps in sensing and NDE technologies are obvious, identifying synergies between technologies and designing effective multi-modal sensing solutions is still an active area where developments are needed. The effective integration of technologies that have a broad range of spatial and temporal resolutions offers an effective means to address the uncertainties and complexities inherent in constructed systems. An especially important issue is the gage-length of strain and/or deformation measurements. For example, measuring live-load strains on a full-scale bridge girder may require at least four-six strain gages with at least a 2 inch gage-length, while trying to capture strains at the tip of a fatigue crack may require one-two strain gages with 1/16-1/8 inch gage-length. The fundamental challenge in experimenting with actual constructed systems is to acquire the most meaningful data, and minimize the uncertainty inherent in the data to facilitate its effective interpretation. This challenge requires more than the minimization of random and bias errors caused by the sensors themselves, which can be mitigated by employing established best practices for both sensor calibration and installation. Rather, overcoming this challenge requires that the epistemic (bias) uncertainty that inevitably results from the complexities of constructed systems (Table 1-1) are recognized, understood, and mitigated to the largest degree possible. Chapter 3 of this report provides a more complete discussion of the key issues related to Step 3 and the case studies provide examples of how such uncertainties have been mitigated during past applications.

1.4.4 Step 4: Data Processing and Feature Extraction

Step 4 of the St-Id process involves the processing and interpretation of data. In general terms data processing activities aim to make the acquired data more appropriate for interpretation. This is typically achieved through cleansing the data of blatant and subtle errors (e.g., spikes, malfunctioning sensors, statistical characterization), improving the quality of the data (e.g., averaging, filtering, windowing, etc.), and then compressing and/or transforming the data to better support interpretation (e.g., the extraction of modal parameters/flexibility, influence coefficients, etc.).

The second stage of Step 4, direct data interpretation, is optional and depends greatly on the objectives and constraints associated with the St-Id application. Direct data interpretation involves fitting mathematical models (also referred to as a non-physics-based model), such as Artificial Neural Networks, Auto-Regressive Models, state space models etc., to the processed data. These models are not formulated with any consideration of the underlying physics of the constructed system, rather they aim to accurately capture and replicate the patterns associated with the data. In this manner, they are most concerned about identifying when the constructed system behavior has changed rather than identifying the underlying cause of the change. This approach has advantages of require minimal user interaction and being able to address large data sets, and, as a result, is a powerful tool for continuous monitoring of structures. However, users should be cautioned that patterns of input and output for constructed systems may greatly change over seconds, minutes, hours, days, months and years based on weather and climate changes. Therefore, assuming a threshold for what constitutes normalcy in data is not possible. Chapter 4 provides a more detailed discussion of Step 4.

1.4.5 Step 5: Selection and Calibration of Physics-Based Models

Step 5 of the St-Id process involves the selection and calibration of physics-based models. These models, in contrast to the non-physics-based models used for direct data interpretation, are formulated to explicitly recognize the underlying physics of the constructed system. If direct data interpretation is employed in the St-Id application, this step may be viewed as optional. However, it is important to note that once a change in response has been identified (using direct data interpretation) it is almost always followed by an investigation into the cause of the change and its influence on the performance of the constructed system. This investigation, which is crucial to informing decisions, requires the use of a physics-based model within the St-Id process.

Although not explicitly addressed by many St-Id applications, the process of model selection is crucial to the overall success. Given the uncertainties identified in Table 1-1, it is clear that simply developing a finite element model with typical engineering assumptions and idealizations may not be sufficient. Rather it is recommended that several different modeling strategies be employed and compared to ensure the model selected for calibration is appropriate. The model calibration process typically involves optimizing a set of model parameters to minimize the difference between the initial model and the experimental results. Approaches to this model calibration (also known as model updating) can be classified based on how they select the parameters to identify, the formulation of their objective functions (to minimize), the optimization approach they employ (e.g., gradient-based or non-gradient-based), and whether or not they explicitly address uncertainties, among others. Chapter 5 discusses the issues related to Step 5 of the St-Id process in greater detail.

1.4.6 Step 6: Utilization of Models for Decision-Making

The ability to utilize the models developed and calibrated (physics-based) or trained (non-physics-based) through the St-Id process for decision-making is essential if the application is to be justified from an economic standpoint. Presently, most reported examples have been in the realm of research, culminating with Step 4 or Step 5. In these applications success has been defined as attaining a good agreement between the measured and simulated properties. Unless this definition of success is expanded to explicitly include the ability of the application to influence the decision-making process, the paradigm of St-Id may never move from the realm of research to widespread applications in practice.

Properly leveraging a calibrated analytical model through scenario analysis, parametric studies, or what-if simulations, in order to influence decisions should be given the same attention and creativity as Steps 1-5. Whether the decision in question is related to improving the performance of a design at different limit-states, or to evaluating the future performances of an existing as-constructed system, influencing decisions is a crucial part of the St-Id process. Without adequate focus on this step, the cost-benefit associated with St-Id will always be unfavorable and applications will only serve to fuel skepticism. In the future, especially as an increasing portion of civil engineering expenditures relate to renewal of existing constructed systems, simulation-based management of our constructed environment will be essential, and will rely on reliable applications of St-Id. Chapter 6 provides a more in-depth discussion of the tools and context related to how St-Id may be used to inform the decision-making process.

1.5 Outline of Report

In forming the ASCE Committee care was exercised to identify the leading experts in the world in each of the steps illustrated in Figure 1-1 and to bring them together so that their expertise may be integrated to advance the engineering and management of civil constructed systems. In this report the state-of-the-art related to each of the six steps is discussed in depth by experts in the corresponding disciplines and application areas, and future research and application needs are be identified (Chapters 2-6).

In addition, this report offers summaries of selected examples of St-Id applications to two most common constructed systems: buildings and bridges. Whether the calibrated models were actually utilized for decision-making and the scenarios and analytical approaches that were employed for this purpose are specifically discussed. The societal systems (policy, planning, financing, legal, etc.) that impact the utilization of St-Id in practice are also identified in each case study along with possible strategies for overcoming current barriers.

1.6 References

Agbabian, M. S., Masri, S. F., Miller, R. K., and Caugher, T. K. (1990). "System identification approach to detection of structural changes." *ASCE Journal of Engineering Mechanics,* 117, 370-390.

Aktan, A. E., and et al. (2007). "Long-term vision for the ASCE technical committee: Structural identification of constructed systems." *SHMII-3: 3rd International Conference on Structural Health Monitoring and Intelligent Infrastructure,* Vancouver, BC.

Aktan, A. E., Farhey, D. N., Helmicki, A. J., Brown, D. L., Hunt, V. J., Lee, K. L., and Levi, A. (1997). "Structural identification for condition assessment: Experimental arts." *Journal of Structural Engineering,* 123(12), 1674-1684.

Aktan, A. E., Helmicki, A. J., and Hunt, V. J. (1998). "Issues in health-monitoring for intelligent infrastructure." *Smart Materials and Structures,* 7(5), 674-692.

Aktan, A. E., and Moon, F. L. (2005). *ASCE-SEI performance of structures track technical committee: Structural identification of constructed systems (2005),* Drexel University, Philadelphia, PA.

Al-Khalidy, A., Noori, M., Hou, Z., Yamamoto, S., Masuda, A., and Sone, A. (1997). "Health monitoring systems of linear structures using wavelet analysis." *International Workshop on Structural Health Monitoring,* Stanford University, Palo Alto, CA.

Andersen, P., and Kirkegaard, P. H. (1998). "Statistical damage detection of civil engineering structures using ARMAV models." *16th International Modal Analysis Conference,* Society of Experimental Mechanics, Santa Barbara, CA.

Beck, J. L. (1990). "Statistical system identification of structures." *Structural safety and reliability,* ASCE, New York, 1395-1402.

Beck, J. L., and Katafygiotis, L. S. (1998). "Updating models and their uncertainties: Bayesian statistical framework." *Journal of Engineering Mechanics,* 124(4), 455-461.

Billington, D. P. (1979). *Robert Maillart's bridges: The art of engineering*, Princeton University Press, Princeton, NJ.

Biswas, M., Pandey, A. K., and Samman, M. M. (1989). "Diagnostic experimental spectral/Modal analysis of a highway bridge." *International Journal of Analytical and Experimental Modal Analysis*, 5(1), 33-42.

Bodeux, J. B., and Golinval, J. C. (2001). "Application of ARMAV models to identification and damage detection of mechanical and civil engineering structures." *Smart Materials and Structures*, 10, 479-489.

Brownjohn, J. M. W. (2003). "Ambient vibration studies for system identification of tall buildings." *Earthquake Engineering and Structural Dynamics*, 32(1), 71-95.

Brownjohn, J. M. W., Dumanoglu, A. A., Sevem, R. T., and Taylor, C. (1987). "Ambient vibration measurements of the Humber suspension bridge and comparison with calculated characteristics." *Institution of Civil Engineers, Part 2*, 83, 561-600.

Brownjohn, J. M. W., Dumanoglu, A. A., and Severn, R. T. (1992). "Ambient vibration survey of the Fatih Sultan Mehmet (second bosporus) suspension bridge." *Earthquake Engineering and Structural Dynamics*, 21, 907-924.

Brownjohn, J. M. W., Moyo, P., Omenzetter, P., and Lu, Y. (2003). "Assessment of highway bridge upgrading by dynamic testing and finite element model updating." *ASCE Journal of Bridge Engineering*, 8(3), 162-172.

Bucher, C., Huth, O., and Macke, M. (2003). "Accuracy of system identification in the presence of random fields." *ICASP 9*, San Francisco, CA.

Chang, C. C., Chang, T. Y. P., and Zhang, Q. W. (2001). "Ambient vibration of long-span cable-stayed bridge." *ASCE Journal of Bridge Engineering*, 6(1), 46-53.

Chassiakos, A. G., Masri, S. F., Smyth, A. W., and Anderson, J. C. (1995). "Adaptive methods for identification of hysteretic structures." *American Control Conference*, Seattle, WA.

Doebling, S. W., Farrar, C. R., and Prime, M. B. (1998). "A summary review of vibration-based damage identification methods." *Shock and Vibration Digest*, 30(2), 91-105.

Douglas, B. M., and Reid, W. H. (1982). "Dynamic tests and system identification of bridges." *ASCE Journal of Structural Engineering*, 108(10), 2295-2312.

Eves, H. (1990). *An introduction to the history of mathematics with cultural connections*, Saunders College Publishing, Fort Worth, TX.

Farrar, C. R., and Doebling, S. W. (1998). "Damage detection II: Field applications to large structures." *Proceedings of NATO Advanced Study Institute on Modal Analysis and Testing,* Sesimbra, Portugal.

Fujino, Y., and Abe, M. (2002). "Vibration-based structural health monitoring of civil infrastructures." *First International Workshop on Structural Health Monitoring,* ISIS, Winnipeg, Canada.

Fujino, Y., Nakamura, S., Shibuya, H., Sato, M., Yanagihara, M., and Sakamoto, Y. (1999). "Forced and ambient vibration tests of Hakucho suspension bridge." *Structures Congress,* ASCE, New Orleans, LA.

Gurley, K., and Kareem, A. (1999). "Applications of wavelet transform in earthquake, wind and ocean engineering." *Engineering Structures,* 21(2), 149-167.

Hart, G. C., and Yao, J. T. P. (1977). "System identification in structural dynamics." *ASCE Journal of Engineering Mechanics,* 103(6), 1089-1104.

Hornbuckle, J., Mercurio, T., Howard, G., Ibanez, P., Smith, C. B., and Vasudevan, R. (1973). "Forced vibration tests and analyses of nuclear power plants." *Nuclear Engineering and Design,* 25(1), 51-93.

Hou, Z. K., Noori, M., and Amand, R. S. (2000). "Wavelet-based approach for structural damage detection." *ASCE Journal of Engineering Mechanics,* 126(7), 677-683.

Huang, N. E., and et al. (1998). "The empirical mode decomposition and the Hilbert spectrum for non-linear and non-stationary time series analysis." *Philosophical Transactions of the Royal Society of London, Series A, Mathematical and Physical Sciences,* 454, 903-995.

Hudson, D. E. (1970). "Dynamic tests of full-scale structures." *Earthquake engineering,* Prentice-Hall, Inc., Englewood Cliffs, NJ, 127-149.

Hudson, D. E. (1976). "Dynamic tests of full-scale structures." *Dynamic Response of Structures: Testing Methods and System Identification,* University of California, Los Angeles, Los Angeles, CA.

Ibanez, P. (1979). *Review of analytical and experimental techniques for improving structural dynamic problems,* 249, Welding Research Council, Shaker Heights, OH.

Jayakumar, P., and Beck, J. L. (1988). "System identification using nonlinear structural models." *Structural safety evaluation based on system identification approaches,* H. G. Natke and J. T. P. Yao, eds., Vieweg-Verlag, Wiesbaden, Germany, 1.

Kapania, R. K., and Park, S. (1997). "Parametric identification of nonlinear structural dynamic systems using time finite element method." *American Institute of Aeronautics and Astronautics,* 35(4)

Kijewski, T., and Kareem, A. (2003). "Wavelet transforms for system identification in civil engineering." *Computer-Aided Civil and Infrastructure Engineering,* 18, 339-355.

Kossiakoff, A., and Sweet, W. N. (2002). *Systems engineering principles and practice,* John Wiley and Sons, New York, NY.

Kou, J. W., and DeWolf, J. T. (1997). "Vibrational behavior of continuous span highway bridge - influencing variables." *ASCE Journal of Structural Engineering,* 123(3), 333-344.

Liu, S. C., and Yao, J. T. P. (1978). "Structural identification concept." *ASCE Journal of the Structural Division,* 104(ST12), 1845-1858.

Maeck, J., and De Roeck, G. (2003). "Damage assessment using vibration analysis on the Z24-bridge." *Mechanical Systems and Signal Processing,* 17(1), 133-142.

Masri, S. F., Nakamura, M., Chassiakos, A. G., and Caughey, T. K. (1996). "Neural network approach to detection of changes in structural parameters." *Journal of Engineering Mechanics,* 122(4), 350-360.

Moon, F., and Aktan, A. E. (2006). "Impacts of epistemic (bias) uncertainty on structural identification of constructed (civil) systems." *Shock and Vibration Digest,* 38(5), 399-420.

Mottershead, J. E., and Friswell, M. I. (1993). "Model updating in structural dynamics: A survey." *Journal of Sound and Vibration,* 167(2), 347-375.

Naghavi, R., and Aktan, A. E. (2003). "Nonlinear behavior of existing heavy-class steel truss bridges." *ASCE Journal of Structural Engineering,* 129(8), 1113-1121.

Nakamura, M., Masri, S. F., Chassiakos, A. G., and Caughey, T. K. (1998). "Method for nonparametric damage detection through the use of neural networks." *Earthquake Engineering and Structural Dynamics,* 27, 997-1010.

Natke, H. G., and Yao, J. T. P. (1986). "Research topics in structural identification." *3rd Conference on Dynamic Response of Structures,* Los Angeles, CA.

Natke, H. G., and Yao, J. T. P. (1989). "System identification methods for fault detection and diagnosis." *Fifth International Conference on Structural Safety and Reliability,* San Francisco, CA.

Shinozuka, M., and Ghanem, R. (1995). "Structural system identification II: Experimental validation." *ASCE Journal of Engineering Mechanics,* 121(2), 265-273.

Smyth, A. W., Masri, S. F., Chassiakos, A. G., and Caughey, T. K. (1999). "Online parametric identification of MDOF nonlinear hysteretic systems." *ASCE Journal of Engineering Mechanics,* 125(2), 133-142.

Sohn, H., Farrar, C. R., Hemez, F. M., Czarnecki, J. J., Shunk, D. D., Stinemates, D. W., and Nadler, B. R. (2004). *A review of structural health monitoring literature: 1996-2001,* LA-13976-MS, Los Alamos National Laboratory, Los Alamos, NM.

Stubbs, N., Kim, J. T., and Topole, K. (1992). "An efficient and robust algorithm for damage localization in offshore platforms." *10th ASCE Structures Congress,* San Antonio, TX.

Teughels, A., and De Roeck, G. (2004). "Structural damage identification of the highway bridge Z24 by FE model updating." *Journal of Sound and Vibration,* 278(3), 589-610.

Vincent, H. T., Hu, S. J., and Hou, Z. (1999). "Damage detection using empirical mode decomposition method and a comparison with wavelet analysis." *2nd International Workshop on Structural Health Monitoring,* Stanford University, Palo Alto, CA.

Wenzel, H., and Pichler, D. (2005). *Ambient vibration monitoring,* John Wiley and Sons, New York, NY.

Yang, J. N., Lei, Y., and Huang, N. E. (2001). "Damage identification of civil engineering structures using Hilbert-Huang transform." *3rd International Workshop on Structural Health Monitoring,* Stanford University, Palo Alto, CA.

Yuen, K. V., and Katafygiotis, L. S. (2002). "Bayesian modal updating using complete input and incomplete response noisy measurements." *Journal of Engineering Mechanics,* 128(3), 340-350.

Zapico, J. L., Gonzalez, M. P., and Worden, K. (2003). "Damage assessment using neural networks." *Mechanical Systems and Signal Processing,* 17(1), 119-125.

Chapter 2
A Priori Modeling

2.1 Introduction

The development of an a priori model within the structural identification (St-Id) process serves to provide estimates of structural responses World Forum on Smart Materials and Smart Structures Technology that will aid in the selection of appropriate experimental approaches and applications (Step 3). The actual modeling approach adopted is dependent on the objectives of the St-Id as well as the complexity of the structure being identified. In some cases simple phenomenological models are sufficient whereas in other cases structural models or high resolution geometric-replica finite element models may be justified. In all cases, the development of a priori models should follow established modeling practices. That is, the effect of modeling assumptions should be examined through the comparison of several models/modeling approaches and the sensitivity of the simulations to parameters with significant uncertainty should be established. In addition to error screening the model, these studies also serve to identify key structural responses and their bounds to ensure the experimental program is robust and reliable. Depending on the objectives of the St-Id, the a priori model may also serve as the model calibrated through parameter identification (Step 5). In these cases, the overall objectives of the study may require more refined a priori models than could be justified to support the experimental program.

2.2 Classification of A Priori Models

Although the term a priori has entered into standard science and engineering vocabulary, general definitions remain informative, such as: "from a general law to a particular instance; valid independently of observation" (dictionary.com); "proceeding from a known or assumed cause to a necessarily related effect; deductive" (American Heritage Dictionary); and "based on hypothesis or theory rather than experiment" (WordNet). Given these definitions, an alternative classification of models (compared to PB vs NPB) that is more relevant to a priori modeling can be put forth; that of predictive versus descriptive.

In general terms PB models can be considered predictive as they rely heavily on the generalized laws of statics, mechanics, dynamics, etc. This basis, which does not require response data from the constructed system, allows such models to be useful in a true a priori sense. This is not to suggest that there is no uncertainty in such models; these issues will be discussed explicitly in the following sections. In contrast, NPB models are descriptive in nature. They are not based on specific generalized laws but are derived principally from various means of data modeling, reduction and interpretation. As such, these models are not appropriate for a priori use. However, once NPB models are trained through the use of response data, they may be considered predictive as they are then capable of estimating future response

STRUCTURAL IDENTIFICATION OF CONSTRUCTED SYSTEMS   19

through forecasting identified patterns and thus identifying when the system has changed. Given this distinction, the remainder of this chapter will focus on PB, predictive techniques as these are the most appropriate a priori models.

2.3 Common Model Types

The most pertinent distinction between the numerous PB modeling approaches for structural identification is that of geometric resolution. The selected a priori model should be commensurate with the uncertainty that prevails as well as the precise motivation(s) for the St-Id application (see Chapter 1 for examples). The resolution and size of the model should be driven by the utility of the St-Id, available information and heuristics about the constructed system, as well as its size, complexity and the experimental resources that are available for Step 3. Most a priori models are based on assumptions of linearity and stationarity. In general these assumptions need not be made; however, in the absence of response data from the specific constructed system, it is difficult to justify the complications associated with nonlinear constitutive relations or stochastic finite element analysis.

Starting with simpler, greatly idealized phenomenological geometric models to help conceptualize a constructed system, together with the site, soil and foundations, and then gradually increasing the detail and complexity of the model as the system is better understood is recommended. Many of the issues associated with a priori modeling are not unique to St-Id applications and different modeling approaches have been developed and discussed. In any case, the utility of the a priori PB models lies in its ability to identify key mechanism and provide an expected range of response to allow an efficient and robust experimental program to be designed and carried out. The following sections provide brief discussions and examples of the most common PB models employed as a priori models.

2.3.1   Phenomenological Models

This class of models has the lowest geometric resolution and typically consists of a

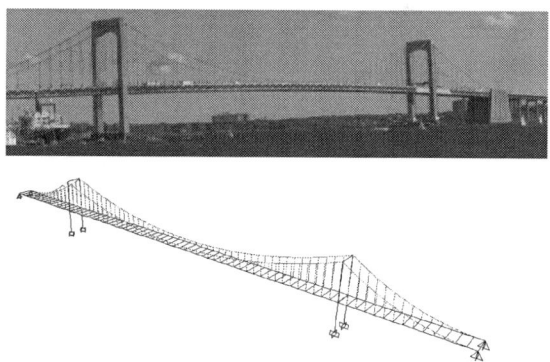

Figure 2-1. Example of a phenomenological model employed for the St-Id of the Throgs Neck Bridge in New York City

20   STRUCTURAL IDENTIFICATION OF CONSTRUCTED SYSTEMS

few elements to describe or investigate the key response mechanisms of constructed systems (Figure 2-1). Although not a strict limitation, these model mostly employ simple one-dimensional (e.g., plane or space frame elements) and discrete elements (e.g., translational or rotational springs, point masses). The primary advantage of this class of model is their transparency to the analyst and computational efficiency. If employed properly, such models can provide great insight into the relative impacts of various global mechanisms of the constructed system in a timely and efficient manner.

2.3.2   Structural Models

Perhaps the most common class of models employed in an a priori manner to support St-Id is the structural model (Figure 2-2). These models typically employ both one-dimensional (plane or space frame elements) and two-dimensional elements (e.g., plate or shell elements). In an effort to remain consistent with the three dimensional geometry of the structure, various link elements, constraints, and rigid offsets are included. The primary advantage of these models is their ability to simulate more detailed, component-level response, and allow the impact of various member-level continuity and boundary conditions on the overall response to be assessed. In addition, they are less dependent than phenomenological models on the understanding of the structural response by the analyst. On the other hand, their construction and error screening is far more time consuming and tedious.

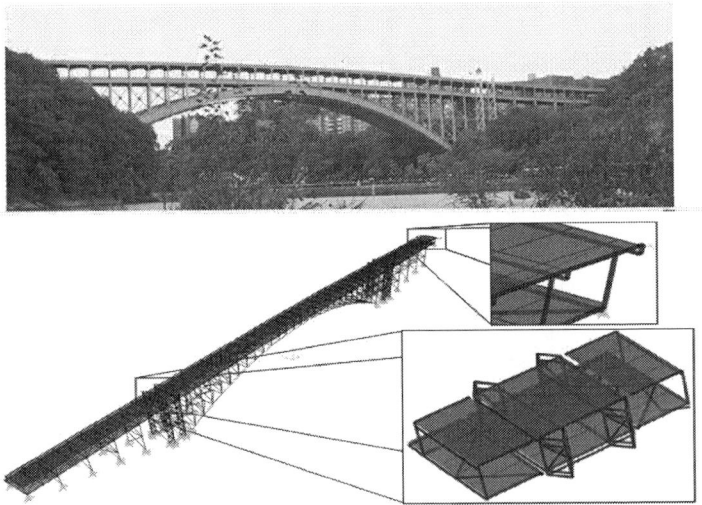

Figure 2-2. Example of a structural model employed for the St-Id of the Henry Hudson Bridge in New York City

2.3.3 Finite Element Models

This class of modeling has the finest geometric resolution, and in some cases may consist of a geometric-replica model (Figure 2-3). As such, these models will employ the full range of finite elements available including three-dimensional solids. The primary advantage of this class of models is their ability to simulate the response and effect of complex structural details, connections, stress concentration, etc. The trade-off of course, is that while these models, in theory, would require less intuition and heuristics related to the response of the constructed system, they are extremely challenging to construct and error screen in a reliable manner. In addition, the computation cost of such models makes them untenable for most constructed systems, and thus their real utility may be as a supplement/complement to a structural model. Although many commercial FE software packages are available to construct such models, they are currently not widely used in civil engineering practice. Very often relatively simple FE codes are interfaced with design guidance to simplify the task for the design engineer. These often have deficiencies for when it comes to structural dynamics. At the other extreme where sophisticated FE codes with pretty interface are used we know there is a tendency to believe the model because it has lots of elements and it looks nice.

2.4 Modeling Constraints Unique to St-Id

While most of the challenges related to the selection and development of a priori models are somewhat universal, there are two unique constraints that must be satisfied for cases where the a priori model will be used for updating. First, the a priori model must be constructed using an analysis package that either incorporates or can interface with updating software if a 'formal' model updating is desired. It is possible to perform a heuristic-based, manual model updating without this constraint; however, this approach can be tedious and is greatly limited. In general, two approaches have been used to satisfy this constraint. First, the use of an analysis package that directly incorporates programming capabilities or the use of third party updating software such as FEMTools or Dakota has been employed. In some cases however, these may be of limit usefulness as such analysis packages may not contain the elements or post-processing capabilities that the analyst desires. The second approach is to export the simulation model to general analysis software such as Matlab, which can then be used to analyze and update the model. The disadvantages of this approach are related to the time and effort required to write the code to solve the model and write an output file that can be read by post-processing software.

The second unique constraint involves the requirement that the most uncertain aspects of the model be parameterized such that they can be included within the updating process. For common modeling approaches this is typically not considered and thus certain modeling habits may require some modifications. A simple example of this issue involves modeling composite action between a slab and beam. For a design exercise it may be most efficient to simulate this using beam and shell elements that are constrained to deform together. In the case of St-Id however, if the degree of composite action is uncertain, then use of constraints are not ideal as they do not provide an updatable parameter related to this behavior. A better approach in

22 STRUCTURAL IDENTIFICATION OF CONSTRUCTED SYSTEMS

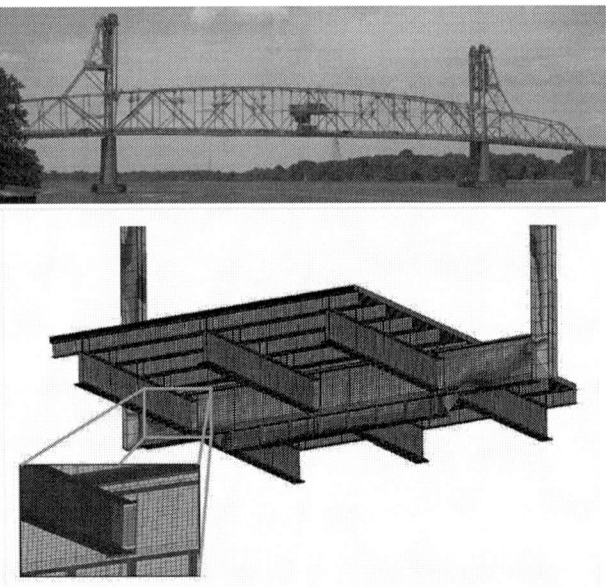

Figure 2-3. Example of a geometric-replica finite element model employed for the St-Id of the Burlington-Bristol Bridge in Burlington, NJ.

this case would be to use link elements and include the stiffness of these elements as parameters within the updating process. Even further, if the model may miss out contributing components such as 'non-structural elements' that may make the crucial difference or make a grossly unrealistic simplification (e.g. the 'spine beam' approach for collapsing torsional and bending properties of complex girders into a simple beam. Without them the best model cannot even get close to a match with test data (Brownjohn et al. 2009).

2.5 Construction of A Priori Models Through 3D CAD

A highly recommended first step for the development of an a priori model is to virtually reconstruct the constructed system by taking advantage of computer aided drafting (CAD) packages. It is not common practice for most engineering offices to leave the 2D tradition of typical design and construction plans. However, there are many risks in trying to construct a computer model for analysis directly from 2D drawings even if up-to-date fabrication and construction plans are available. In the case of many major structures, only a virtual reconstruction by 3D CAD modeling would definitively reveal any lack of information or details or any inconsistencies in an existing set of 2D plans. More importantly, a 3D CAD representation offers a means of physically conceptualizing the structure (Figure 2-4a) and can be independently checked through overlaying photographs (Figure 2-4b). It is common that many critical details of a large and complex constructed system, especially at connections, interfaces and supports, may differ between construction drawings and 'as-built'. A 3D CAD based on photographs and in-situ measurements would serve to confirm the as-constructed configuration.

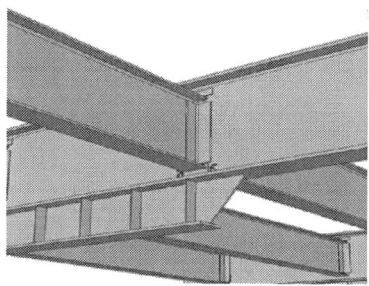 

Figure 2-4. (a) Example 3D CAD of a constructed system and (b) photo overlays to error screen overall geometry

It follows that trying to construct a model for a bridge without actually going to its site and inspecting it should not be acceptable, yet it is seen in several cases with generic and overly-idealized models without incorporating any information from site-visits or even from field inspection reports. Using close-range photogrammetry and reverse-CAD, it is possible to capture as-is details of a bridge and also check its dimensions. With the advent of digital cameras and easy-to-operate software, reverse-3D CAD of at least some of the critical details of a bridge may be constructed in a feasible manner.

2.6 Quality Control Requirements

In cases where critical parameters are highly uncertain or where a structure may exhibit very complex responses, simple trial experiments may be conducted prior to a commitment to a specific a priori model. These trial experiments may typically employ only a few, roving sensors and may be used to approximate the fundamental frequencies of the structure. This information is then used to aid in the selection of an appropriate a priori model and for designing the complete experiment (Step 3). The goal of these initial experiments is to ensure that the a priori models are sufficient to support the more comprehensive experimental program – not to replace it.

The quality control requirements for finite element analysis are discussed in the following sections and are broadly classified as:

- Quantifying modeling errors, especially conceptual errors and incompleteness
- Quantifying input errors

- Quantifying errors that may occur during analysis

2.6.1 Modeling Errors

The first category of errors are the most difficult to detect, especially in the case of large and complex structures. In this regard, the most common errors arise due to over-idealizations of the geometry, boundary and intermediate support conditions, connection stiffnesses, member releases, interfaces between various structural sub-systems, movement systems, etc. The general rule is to carefully evaluate the relative contributions of all possible displacement and deformation mechanisms and incorporate these in the model unless it is justifiable to ignore some. A common example relates to the importance of axial, shear and torsional deformations and any effects of geometric nonlinearity on local and global response, especially when creep and temperature effects are considered. When structural elements of complex cross sections, such as compound tubular sections are used in conjunction with post-tensioned construction, the experience required for accurate and complete modeling becomes significant. Complete and proper modeling of intended or unintended movement mechanisms is a common difficulty. In some cases, the friction in intended movement mechanisms may lead to a locking of these mechanisms. The experience of the engineer constructing the model, and the ability of the analyst to identify the physical implications of any discrepancies between experiment and analysis are key factors for detecting and eliminating the conceptual errors in a model.

2.6.2 Input Errors

Related to the second category, an important issue as the size of a model gets larger is the need for quality control to assure the accuracy of the input data. Generation of a 3D CAD model that is then scrupulously and systematically checked for geometric accuracy of all local and global details against 2D fabrication and construction drawings and up-to-date photographs, which may then be directly transferred to a computer model, is a good measure for quality control. For example, Catbas et al. (2007) developed a structural model of a long-span truss bridge that incorporated 7,150 nodes, over 43,000 degrees of freedom, 8,574 space frame elements and 2,890 shell elements. Although this model was constructed by transferring a 3D CAD model directly as an input file to an analysis program, various element data categories were subsequently discovered to contain errors that were not discovered by checking the output of static or eigenvalue analysis. However, these errors did lead to significant errors in some of the member forces and were discovered only after extensive checking of the input data through transferring the data files into spreadsheet format. Only by re-arranging member properties in the spreadsheet was it possible to detect and correct the erroneous entries.

2.6.3 Analysis Errors

The third category of errors relate to those that occur during an analysis, and in general would not be discovered unless analysis output is suspected. Until all of the global and local displacements, reactions, forces and stresses are verified for physical consistency and correlated against a sufficient amount of reliable experimental data including frequencies, mode shapes, displacements, rotations and strains, such errors should always be considered. It is especially important to validate finite element models against a spectrum of multi-modal global and local response measurements and not just one kind of response such as accelerations or strains.

The fundamental principles that guide testing the reliability of analysis results in the case of phenomenological models remain valid for checking the reliability of analysis using finite element models. However, while these checks are necessary, they are not sufficient in the case of finite element models. For example, verifying that the external equilibrium is maintained between the applied loads and the support reactions is not adequate for assuring the reliability of local stress output from a large finite element model. In some cases, even if there are no errors in the input data, the mathematical and extrapolation formulations of the finite elements used and their geometry may lead to significant errors in the resulting displacements and stresses without affecting global equilibrium. An example is the "locking" phenomenon associated with certain finite elements resulting in a finite element model having an apparent stiffness that may be significantly higher than the structure being simulated.

Another example relates to apparent nodal stresses that may greatly exceed the actual average stresses within an element due to numerical errors. Finally, even slight mechanisms of nonlinearity such as opening of previously formed cracks or slippages at connections in the actual structure may contribute to significant attenuations in peak stresses that may not be properly simulated in the analysis. A recommended approach for the engineer is to gain experience with all aspects of the software and its finite element library by conducting analyses of benchmark problems such as a deep beam or a simply-supported plate under distributed loading and to correlate the numerical results of the analysis with those from the "exact" theoretical results.

2.7 Closing Remarks

The development and use of an a priori model is a critical step within the St-Id process. It serves to help conceptualize a structure, identify key responses (and their bounds) and aid in the selection of appropriate and robust experimental approaches (Step 3). While there are several a priori modeling approaches that may be appropriate (depending on the objectives of the St-Id as well as the complexity of the structure), the usefulness of any approach is typically governed by the care and experience of the engineer.

2.8 References

Brownjohn, J. M. W., Pan, T. C., and Deng, X. Y. (2009). "Macro-updating of finite element modeling for core systems of tall buildings." Proceedings of the 14th Engineering Mechanics Conference, ASCE, Austin, TX.

Catbas, F. N., Ciloglu, S. K., Hasancebi, O., Grimmelsman, K. A., and Aktan, A. E. (2007). "Limitations in structural identification of large constructed structures." *Journal of Structural Engineering,* 133(8), 1051-1066.

Chapter 3
Experimental Considerations

3.1 Introduction

A fundamental component of the Structural Identification (St-Id) process is the experimental process leading to 'data' in various forms and at various levels of refinement, that are used in the analysis tools to decode the performance of a structure. Hence this section of the report concerns the experimental aspects of full-scale investigations of civil infrastructure for St-Id. St-Id goes beyond System Identification in a parametric domain, to provide specific information about the structure and its performance. Simply stated, St-Id uses the results from any number of static and dynamic measurements as a first step towards developing more reliable conceptual or numerical (finite element) models. These models are used to evaluate and predict in-service structural performance, and to support operational and maintenance decisions. Unfortunately, full-scale evaluation of in-service structures is typically limited to the low-level response regime of the structure because of difficulties exciting these structures in a controlled manner at higher levels necessary to validate performance at the limit states of design. As in-service structural performance can be characterized in terms of both static and dynamic parameters, the experimental part of the St-Id process is likely to involve measurements of both dynamic and static signals. Note that while all signals vary in time, dynamic signals used for System Identification are defined as varying fast enough (about an equilibrium configuration that may itself vary more slowly in time) that inertia (mass) properties of the structure are engaged.

The focus of this chapter is thus on the experimental considerations leading up to the delivery of data for Structural Identification purposes. All of the related aspects, including experiment design and execution, sensors and data acquisition, and data storage and transmission are addressed in following sections. While the focus of St-Id is increasingly (and correctly) shifting to data interpretation, the success of that process relies on the quality and quantity of the input data. Experimental data may be readily processed and reprocessed for St-Id in many different ways in a comfortable office environment; however, the difficulties and restrictions associated with acquiring experimental data from full-sale structures require extensive and unusual levels of experience and experimental capabilities. For instance, experimental data often must be successfully collected in a single shot opportunity under challenging field conditions and with stringent time constraints. This chapter is aimed at researchers and practicing engineers who wish to understand and add to the wealth of full-scale performance data that is needed to improve the practice of building and maintaining civil infrastructures.

Full-scale implementations of St-Id experiments for constructed systems have received a number of reviews in the past. These have mainly focused on dynamic evaluations (Eyre and Tilly 1977; Hudson 1970; Hudson 1976; Ibanez 1979; Rainer

1979; Schiff 1972; Srinivasan 1984), while a more recent review (Brownjohn 2007b) covers the broader area of full-scale monitoring including static effects. This is a new and updated contribution that focuses on experimental aspects. Relevant material from the earlier St-Id review commissioned by ASCE group (and having considerable input from the Los Alamos structural health monitoring team) provides a very comprehensive and complementary treatment description of the experimental aspects, biased towards structural health monitoring applications (Sohn et al. 2004).

The design of an experimental program for characterizing and evaluating full-scale performance of civil infrastructure is governed by a large number of constraints including:

- Expectations from the experimental studies, requirements, constraints and specifications (whether commercial, applied research or pure research, there is always a client);
- Experiment duration (different strategies due to robustness, cost and manpower constraints are adopted according to the type of exercise which broadly fits in three categories);
- Structural attributes that will affect the quantity and types of instrumentation used;
- Environmental and operational constraints;
- Logistics and accessibility.

Experiment design may be aided by other information that can include observations and preliminary measurements recorded during reconnaissance visits, finite element simulations from the structural engineer and other background information including structural/site drawings and maintenance and inspection reports. Techniques for optimizing sensor placement (Worden and Burrows 2001) and (Sanayei and Javdekar 2002). are available to assist experiment design, including procedures for virtual modal testing (Ewins 1999).

The ultimate success of any experiment will be heavily influenced by the experience, style and preferences of the experimental specialists, operating within the various constraints. Careful design can be formalized through written method statements which will inform the client about the experiment and guide the execution of the site work, while also addressing safety requirements and risk assessment. Clear statements of specific deliverables and their format will guide the execution of the whole procedure.

This section focuses on the on-site experimental procedures. In some cases, such as condition assessment exercises (e.g., modal tests), on-site data analysis is essential for ensuring the quality of the measurement data, and these steps cannot be divorced from the hardware and software related aspects of the experiment. Data processing and analysis procedures are, however, described in a later chapter.

The experimental procedure can be simply and roughly divided into a number of components that, while they affect each other in the practice, are considered separately in this section. The main components of a St-Id experiment are thus:

1. Selection of inputs, their locations and their means of measurement.
2. Selections of outputs, their locations and their means of measurement.
3. Gathering and transmission of signals to recorder/logger.
4. Conversion of signals to storable form, typically via analog-to-digital conversion for storage on computer disk digital streaming.
5. Data storage, typically on local or remote computer, occasionally on a dedicated logger.
6. Data inspection/quality control by real-time local processing/presentation (details in a later section).

A seventh component can be argued for monitoring applications, and while it should not be neglected is beyond the scope of this chapter:

7. User interface and real-time alerting/reporting

While details of installation procedures (e.g., type of cable, water resistance ratings, cable routing etc.) may be critical for each study, they are too specific to be discussed here, as they depend on an even bigger range of practical constraints.

3.2 Classification of Experiments based on Input

It is possible to characterize full-scale experimental investigations according to the type of input or loading as follows: (1) whether inputs are dynamic or static (i.e., according to whether or not they engage inertial effects), (2) whether the inputs are controllable, and (3) whether the inputs are measurable. Un-measurable and uncontrollable inputs used in isolation constitute what is termed 'ambient' inputs, otherwise they amount to noise when controllable and measurable inputs are used. The following nomenclature may be used to classify the type of input used in conjunction with a full-scale St-Id experiment:

- Static Input
  - Controllable (measurable and un-measurable) static loads (3.2.2.1)
  - Uncontrollable (measurable and un-measurable) static loads (3.2.1.2)
- Dynamic Input
  - Controllable (measurable and un-measurable) dynamic loads (3.2.2.1)
  - Uncontrollable measurable dynamic loads (3.2.2.2)
  - Uncontrollable and un-measurable dynamic input (ambient dynamic excitation) (3.2.2.3)

3.2.1 Static Input

3.2.1.1 Controllable (measurable and un-measurable) static loads

Either trucks or other load sources such as concrete blocks or water-containers may be used to load a bridge while critical responses are measured. The merits of this test technique depend on the quantity and reliability of instrumentation and data acquisition. In general, it is not recommended to conduct such a test unless a reasonable amount of instrumentation is to be utilized. This test can provide information to be interpreted as well as it can be used as a complement to dynamic tests, to provide a reality check and to provide a closer insight into local response mechanisms which modal analysis could not provide.

Such inputs are relatively rare for full-scale experiments on real structures because of the scale of the load required to generate measurable structural responses. A relatively common example is load testing (diagnostic and proof testing) of bridges that often involves the use of heavy vehicles, either stationary or moving (Calcada et al. 2005; Marecos et al. 1969; OECD 1998), with occasional examples of static testing to destruction (Haritos et al. 2000). Load testing of roofs or foundation piles may also be performed by applying water balloons or kentledge.

3.2.1.2 Uncontrollable (measurable and un-measurable) static loads

This category covers a large range of structural monitoring programs that generally include elements of dynamic load and response monitoring, particularly in the case of traffic, wind and temperature which generate quasi-static and dynamic response. However, these examples are the more holistic exercises studying a wide range of external influences and corresponding structural responses.

Structural monitoring applications predominate in bridges so that most of the high-profile exercises concern long span structures, although there is also a history of research on long and short-term monitoring, both static and dynamic, for assessment and management of short span bridges in Canada (Bakht and Jaeger 1990) and USA (Yanev 2003), and for evaluation of new construction technology (Bell et al. 2008).

3.2.2 Dynamic Input

3.2.2.1 Controllable (measurable and un-measurable) dynamic loads

This type of test is generically defined as a forced vibration test (FVT). It remains the most popular method of testing for automotive and aerospace structures and was historically the preferred method for civil infrastructure because of the advantages of a known and controllable input. The reasons why artificial forcing was and is used are very relevant for this report, as are the reasons why technology for St-Id using vibration response data alone has become a very powerful tool and remains a major component of long term performance studies. European researchers in particular are now switching from FVT to alternative procedures such as 'free vibration' and ambient vibration testing (AVT), (Cunha et al. 2006; Cunha et al. 2007) but there remain specific circumstances when FVT is the only viable solution, such as floor vibration performance assessment requiring reliable experimental assessment of modal mass. Transfer functions or frequency response functions (FRFs) scale the input (forcing) to output (response) via either mass or stiffness so both can be identified using this type of test. Stiffness information with good signal-to-noise ratio can be recovered using modest forces due to resonant amplification, and controllable dynamic loads provide the possibility of studying nonlinearities (Jeary and Ellis 1981).

Numerical studies using dynamic measurements have been conducted by several researchers. Recently, a method is developed for structural mass and stiffness estimation including damping effects and using vibration data (Esfandiari et al. 2009). It uses the frequency response function (FRF) data for finite element model updating via a quasi-linear sensitivity equation of structural response. The change of mode shapes is expressed using modal expansion in frequency domain. The FRF data are compiled using the measured displacement, velocity or acceleration of the damaged structure. A least-square algorithm method with appropriate normalization is used for solving the system of equations with noise-polluted data. Sensitivity equation normalization and proper selection of measured frequency points improved the accuracy and convergence in finite element model updating. This methods shows that it can detect, locate and quantify the severity of damage within structures. A decomposed form of the FRF was used in (Esfandiari et. al 2010) as an alternative to (Esfandiari et al. 2009). The change of mode shapes is expressed as a linear combination of original eigenvectors of the intact structures. Results of a numerical truss model show the ability of this method to identify location and severity of damage at the elemental level in a structure. A similar method is also developed to detect changes in stiffness and mass parameters of a structure utilizing strain data in the frequency domain (Esfandiari et al. 2010). This method was successfully applied to the simulations of a plane truss and a plane frame structure using strain data. For this family of finite element methods using FRF data and a quasi-linear sensitivity equation exhibited a fast rate of convergence even using a subset of the measured response. Continuation of this work includes use of laboratory structural model testing and full scale bridge test data.

FVT may also be necessary when the available ambient excitation is unable to excite critical structural modes with a good enough signal-to-noise ratio for reliable measurement. Hence in some structures (e.g., concrete arch dams and football stadia (Reynolds et al. 2007)) FVT and AVT may *both* be required to reach a full understanding of the structure. FVT can be subdivided into the means of applying the force and the type of structure tested.

3.2.2.1.1 Controlled Traffic

A crawl test may have distinct advantages over a stationary load test from a traffic-control viewpoint. A numerical optimal polynomial decomposition technique was developed (Carne and Stasiunas 2006) to extract the influence line for a response which is measured under a slow-moving truck, as long as the weight of each axle is known. In the case of bridges with WIM scales this information will be available for all traffic. The test truck should follow the same path along the bridge every time, and the dynamic components of response should be discernable from the static component. Controlled load tests on long span bridges require additional requirements in terms of load levels, loading patterns and testing constraints. There are a few examples of load tests on long span bridges (Catbas et al. 2007).

3.2.2.1.2 Rotating eccentric mass exciters

This is the classical form of FVT and has a very long history, [see Hudson (1964) for one of the first descriptions of shakers used by CalTech]. Simple forms of rotating mass exciter for exciting full-scale civil structures can be created using industrial vibration generation devices (e.g., for sieves), but purpose-built devices for structural testing are still in use and being manufactured (e.g., by ANCO Engineers). The technology is well described in Severn et al. (1980); rotating eccentric mass (REM) exciters use a pair of contra-rotating masses to generate a uni-directional force that is proportional to the angular velocity ω (squared), the magnitude of the total eccentric mass m and its radius of gyration r. Two fundamental limitations of a REM exciter are the bearing capacity and that the created excitation is sinusoidal.

3.2.2.1.3 Linear or reciprocating mass exciters

There is a range of possible forms for such devices, but for civil structures they usually involve a hydraulic actuator driving a movable mass. Electro-dynamic shakers usually have smaller force capacities and are more popular for automotive and aerospace applications, however' long-stroke' electro-dynamic shakers, where stroke length allows the shaker force to be developed from low frequencies are now widely used in civil applications. While their peak output (450N) is less than the multi-kN outputs of hydraulic shakers, they do not have the logistical disadvantage of requiring hydraulic power packs. While small shakers may not be capable of global structural excitation, they may be useful for local measurements e.g., of stay cables in bridges or floor panels in a building. As with REM exciters the harmonic (sinusoidal) force output is given by mlw^2 where l refers to the stroke length of the armature. Figure 3-1 shows electro-dynamic shakers configured for vertical excitation and for horizontal excitation with enhanced low-frequency performance (due to extra weights). These shakers are not limited to harmonic forcing but can reproduce and generate (subject to the system transfer function) a forcing function covering a broad frequency band.

Examples of signal types available for linear shakers, generally provided by proprietary spectrum analyzers, include 'chirp' signal (a sinusoid whose frequency is swept between limits), pseudo random binary signal (PRBS), true random, burst random and of course (stepped) sine. Additionally, multiple shakers distributed over a large span structure can be used to distribute excitation forces using uncorrelated

Figure 3-1. Electro-dynamic shakers in horizontal and vertical mode.

random excitation so that multiple columns of the frequency response function –FRF– matrix can be populated. Alternately they can be arranged with appropriate phases and amplitudes for 'normal mode' testing to excite close and complex modes. This type of testing is referred to as multiple input multiple output (MIMO) and is popular in the aerospace industry especially for aircraft ground vibration testing where it has evolved from stepped sine normal mode testing (Anderson and Mills 1971) to more efficient methods using optimized signals such as burst random (Hutin 2000).

3.2.2.1.4 Transient (impulsive or impact) testing

Impulsive testing is an attractive proposition for full-scale dynamic testing because the short duration of the excitation translates to a broad-band excitation. Typically no main power supply is needed so the technique does not have the logistical constraints associated shaker-based FVT. The advantages are simplicity, logistics and bandwidth of excitation. Disadvantages include the need for acquisition systems with high dynamic range to capture both the initial large amplitude response and the tail of the decay as it reduces to instrumentation noise floor levels. Some of the tradeoffs between shaker and hammer testing are discussed in Reynolds and Pavic (2000) and examples where comparisons are made with other techniques include the Z24 exercise (Kramer and De Smet 1999) where calibrated drop weight, shaker and ambient testing were used. The data from the Z24 tests were used in a benchmark test of system identification procedures reported by Peeters and Ventura (2003) including one comparing results from all three methods (Luscher et al. 2001).

3.2.2.1.5 Impact hammer

Instrumented hammers are a portable, easy to use and versatile tool for conducting forced vibration testing. This method has been used in a range of applications, although typically with smaller structures. Because impact hammers are highly portable, they permit the operator to rove the excitation location to populate more columns of the experimental frequency response function (FRF) matrix –effectively MIMO testing- and to carry out reciprocity checks, with some limitations (Avitabile 1998).

3.2.2.1.6 Drop-weight

There are several limitations associated with the use of an instrumented hammer for testing; firstly the peak force is limited, secondly it takes skill and experience to deliver repeatable and good quality impacts with no double hits, transducer overload, etc. Thirdly the hammer operator usually has to stand on the structure and could affect the structure's dynamic characteristics because of the human dynamic properties (mass, stiffness, damping). Such effects have been clearly demonstrated on pedestrian and assembly structures (Dougill et al. 2006). Hence drop-weight systems which deliver large-scale repeatable impulses with minimal effect on the structure dynamics are popular, particularly for bridge or floor tests (since it is easier to rig a vertical impacting system). Load cells may or may not be installed, and with variable height of the drop weight it is possible to study structural linearity (Green and Cebon 1994). There are examples of drop-weight with vehicle impact as shown in Figure 3-2.

3.2.2.1.7 Snap-back, step relaxation or free vibration

This is strictly not an excitation so much as it is an initial condition of displacement imposed on a structure by a very large static force that first has to be provided and then safely and suddenly released. This is usually accomplished via cables and some kind of structural fuse such as an explosive bolt cutter.

Comparison has been made with ambient vibration tests in several cases e.g., Gentile and Cabrera (1997); Ventura et al. (1996), where, in general, good agreement was obtained between dynamic properties identified using both excitation methods. One advantage of step relaxation is that it engages large amplitude response so that non-linear effects (e.g., the variation of frequency with amplitude) can be investigated. A disadvantage is that the strength of modes involved in the response depends on the relative contribution to the static deformed shape, which generally resembles most strongly the fundamental mode. Additionally, greater emphasis on health and safety considerations tends to rule against use of such techniques.

3.2.2.1.8 Other forms of controlled excitation

As well those categorized above, a range of exotic and imaginative methods have been used for dynamic testing such as pulse-train generation (Safford and Masri 1981), using a swinging crane to excite a building (Glanville et al. 1996) or a bridge (Stiemer et al. 1988; Talbot and Stoyanoff 2005), using a swinging bell to excite a cathedral (Patron-Solares et al. 2005), jumping on a force plate (Brownjohn and Tao 2005), vehicles (Buckland et al. 1979; Calcada et al. 2005; Lee et al. 1987), and PZT patches (Park and Inman 2007) for low level local excitation (see section 3.3.9 for use of PZT patches as sensors).

3.2.2.2 Uncontrollable measurable dynamic loads: seismic excitation

Seismic excitation can be considered as the only form of uncontrollable but fully measurable dynamic loading (wind and wave forces cannot be measured absolutely). Seismic excitation of a structure at its supports provides a body force in the direction of the shaking so if the ground motion can be characterized, then the structural system can be identified, although only mass distribution rather than the absolute mass can be determined. There are numerous cases where the dynamic response of a structure is excited by strong motion or micro-tremors, but if the ground motion is not recorded, such measurements are a variety of AVT.

Several studies make use of seismically-induced motions to generate responses used for AVT, without necessarily making specific use of ground motion measurements. One of the earliest recorded exercises, by the US Coast and Geodetic Survey (United States Coastguard and Geodetic Survey 1936) used micro-tremor excitation of buildings. More recently Tanaka et al. (1969) provides one of many examples making use of unmeasured seismic excitation.

For bridges, Higashihara et al. (1987) describes a test of a suspension bridge anchorage using both seismic motions and micro-tremors, and Wemer et al. (1987) describes an instrumented two-span concrete bridge subjected to a strong motion earthquake.

34 STRUCTURAL IDENTIFICATION OF CONSTRUCTED SYSTEMS

Figure 3-2. Vehicle impact (Annacis Bridge, Vancouver)

3.2.2.3 Uncontrollable un-measurable dynamic input (ambient dynamic excitation)

As technology for AVT develops, a higher proportion of full-scale tests use the technique. Reviews of AVT are included among the earlier cited review papers, but there are a few additional reviews specifically for AVT (Udwadia and Trifunac 1974) which cover all types of civil structures and provide details of an instrumented building in California.

'Operational modal analysis' (OMA) is the 'output-only' equivalent of experimental modal analysis (EMA) where the experimental component is the AVT or ambient vibration survey (AVS). AVT procedures have been in use for a century or more, and it is the recent developments in system identification theory described in later parts of this report that have led to its increased use for St-Id. Techniques such as frequency domain decomposition, stochastic subspace identification and other analysis tools are now being routinely employed in conjunction with AVT and they are discussed in a separate section. There are numerous examples where the results of AVT and FVT are compared for buildings, bridges and offshore installations (Kramer and De Smet 1999; Rubin 1980; Trifunac 1972).

3.2.2.3.1 Wind

Wind excitation is typically the dominant loading on low frequency structures, i.e., those with fundamental frequencies below 1Hz, with diminishing effect at higher frequencies. This includes long span bridges and tall buildings. Alongwind loading is the nearest excitation source to the ideal stationary Gaussian white noise process, exhibiting excellent random character and relatively smooth spectra over a broad frequency range. There is also relatively weak correlation over the span of the type of structure likely to be excited; hence in principle both symmetric and non-symmetric modes may be excited. There are many applications of AVT using wind and some of them are discussed in the later parts of this text.

3.2.2.3.2 Traffic

Full-scale testing on a highway bridges can rarely be undertaken in the absence of traffic, unless the bridge is just about to be opened (Brownjohn et al. 1992) (or it has been decommissioned, as for Z24), hence traffic excitation is ever-present. Alas, unlike wind, dynamic excitation due to highway vehicles is far from white noise, but is colored, with concentration of force around 'body bounce' and 'axle hop' frequency bands of approximately 2-5Hz and 10-15Hz (Cebon 1993), so that it may not be possible to identify a peak in a Fourier spectrum uniquely as a bridge mode. Further, vehicle traffic is not usually a constant stream (i.e., it is non-stationary) and for small bridges, the vehicle dynamics can have a significant influence on modal properties, although cars (automobiles) have minimal effect (Brownjohn et al. 2003). Very recently, researchers proposed and demonstrated load tests using operating traffic monitored by means of synchronized sensing and video streaming. This approach does not require any bridge closure, is very practical and cost effective (Zaurin and Catbas 2010, 2011).

3.2.2.3.3 Waves

Wave motion has been used extensively by the offshore oil industry as an ambient source of dynamic excitation. A primary drawback of wave excitation is that it tends to excite only the lower frequency modes of the structure; typically, the wave motion has most of its energy content below 2 Hz. As with many other ambient vibration sources the wave motion is accompanied by many other vibration sources from equipment operating on the oil platforms e.g., Spidsoe and Hilmarsen (1983), so that identification (in the presence of harmonic excitation) is a major challenge, only recently being addressed (Peeters et al. 2007).

3.2.2.3.4 Pedestrians and crowds (without prompting)

Increasingly, pedestrians are used for AVT of bridges as well as for proof testing (Brownjohn, Fok, Roche, and Omenzetter 2004; Caetano et al. 2007; Dziuba et al. 2001; Fitzpatrick et al. 2001). As with vehicular traffic, pedestrians provide a non-stationary colored excitation (Brownjohn et al. 2004) that is accompanied by interaction with the structure which is less of an issue for heavier bridges, e.g., Changi Mezzanine Bridge, Singapore (Brownjohn, Fok, Roche, and Moyo 2004; Brownjohn, Fok, Roche, and Omenzetter 2004). Similarly, excitation by humans is the major cause of excessive vibrations at football stadia (Pernica 1983) and can be used to track modal properties as a means to back-analyze the load (Reynolds and Pavic 2006).

3.3 Sensors and Sensor Classification

Sensors are one of the most critical components of Structural Identification since the quality of the analysis results directly depends on the quality of the data collected. There are many text books that give very detailed summaries of the various sensors used for measurement systems (Doebelin 1990; Dunnicliff 1994; Huston 2011; McConnell 1995; Miller et al. 1992; Reese and Kawahara 1993).

Sensors can be categorized according to measurand or operating principle. For example a vibrating wire sensing element (Yu and Gupta 2005) features in a wide range of static instrumentation, while Fiber Bragg Grating (FBG) arrays can be used to measure temperature, strain, pressure, acceleration, etc. On the other hand, acceleration signals can be recorded by FBG sensors, piezo-electric sensors, force-balance servo-accelerometers, etc.

Selection of the sensors, calibration and installation technique, signal conditioning and data acquisition is a challenging design problem. Instrumentation for field testing is in general far more challenging then instrumentation in a laboratory. Since sensors are typically chosen according to the measurand, this is the more logical categorization. The sensors are categorized according to the measurands in the following sections.

3.3.1 Acceleration

Accelerometers are the default choice for short-term dynamic measurement applications and their versatility makes them popular for long-term monitoring. Currently, there is a wide range of different accelerometer types available; however, many types are not suitable for field structural applications. The main variants of accelerometer types include:

- Servo-accelerometer
- Piezo-electric accelerometer
- Capacitive accelerometer
- Strain gauge accelerometer
- MEMS (capacitive and piezo-resistive) accelerometer (Partridge and Kenny 2000)
- FBG accelerometer (Todd et al. 1998)
- Laser vibrometer (Rossi et al. 2002)

Accelerometer characteristics that affect their suitability for civil infrastructure applications are cost, dynamic range, resolution, noise floor (Brownjohn 2007a), frequency range (usually DC to sub kHz), power consumption, cabling requirements and limitations, and conditioning requirements. Experience has shown that high specification units repay the investment of higher cost, that accelerometers commonly used for aerospace/automotive application do not perform well in civil structure environments and that instrumentation contractors often lack an understanding of the challenging environments where low frequency, low level vibrations that need to be measured, even with explicit instrumentation specifications. Advances in MEMS technology are closing the gap with traditional technologies, to provide low-cost units with sensitivity as good as 10mg, which is suitable for civil applications.

3.3.2 Displacement

Structures move dynamically (i.e., engaging inertial effects) due to wind, seismic, vehicular and even pedestrian loading, and they also move at sub-dynamic rates due to thermal effects, settlement, creep and variation of static loading. Hence structural movements are necessarily time-varying to some degree and comprise both dynamic quasi-static components. With the exception of sub-dynamic tectonic effects,

STRUCTURAL IDENTIFICATION OF CONSTRUCTED SYSTEMS   37

movements, which are both translational and rotational displacements, are necessarily deflections from a 'reference state' that result in deformation of the structure. Displacements are hardly absolute values (even when using technology such as GPS) and invariably require definition of a reference datum, hence in reality it is deformation that is being measured as deflections at sample locations. Surveying techniques have generally been used for measurements of static position, but the dividing line now becomes blurred as total stations and GPS receivers are now also used for structural deformation measurements.

Given that displacements are relative, it is usually enough to determine displacements with respect to an unloaded, un-deformed state. Dynamic displacements related to vibrations rather than quasi-static effects can be recovered from accelerations by double integration after high-pass filtering, but the lost low-frequency components have great value in studies of wind and thermal effects, for example, and their recovery is discussed later.

Recently, a number of studies of tall buildings (Brownjohn et al. 2005; Dalgliesh and Rainer 1978; Kijewski-Correa, Kilpatrick et al. 2006; Littler and Ellis 2007) have been conducted principally to study wind effects in relation to loading code development. Measurement techniques in these cases included conventional survey techniques, lasers, plumb lines, accelerometers and GPS.

3.3.2.1 Laser and LED devices

Several laser-based displacement measurement technologies are described in (Bougard and Ellis 2000). These technologies include the laser interferometry for measuring motion along line of sight at distances up to 500m and down to frequencies of 0.1Hz, a VHS laser for short-range high speed (e.g., blast) deflection measurements, a 3D system for static tracking of multiple locations in 3 directions and tracking lasers which measure angular deflection of a target by tracking relative motion of the laser beam fired from a remote position in the structure.

More recent forms of this technology have been commercialized for applications to measure dynamic transverse motion as distances over 100m (Ahola and Tervaskanto 1991; Myrvoll et al. 1994). These systems require a reflective marker, resolution is about 0.01% of range and frequency response is compatible with structural vibration modes. Smaller versions of the sensor have been used in non-contacting displacement measurements of wind-tunnel section models (Zhang and Brownjohn 2004) in parallel with laboratory lasers.

While designed for automotive and aerospace applications, Laser Doppler Vibrometers (LDVM) have been used for bridge dynamic testing using both displacement and velocity measurements, and more recently for acceleration (Siringoringo and Fujino 2006). Since they use a single laser beam, LDVMs cannot measure multiple locations simultaneously so their applications for modal testing of civil infrastructure are rather limited, and they are not used for long-term monitoring.

3.3.2.2 Image tracking via CCD arrays

There is a small but growing number of image tracking applications in civil St-Id (Caetano, Silva et al. 2007) one of the earliest forms being the 'optometer' developed by ISMES, Italy (Zasso et al. 1993). The system was used to track movements of Humber Bridge in 1990. Figure 3-3 shows the horizontal and vertical targets with alternate black and white bands were fixed to inspection gantries at quarter span (350m range) and midspan (700m). 2000mm lenses fixed on a concrete plinth in a hut at the north (Hessle) tower pier were used to focus target images on CCD arrays from which level thresholds were used to locate the angle to the line of sight. Resolution depended on range, at 100m range it was 0.5mm, and frequency response dropped off rapidly above 0.3Hz.

In parallel to the optometer, a more sophisticated system was deployed by Bristol University. A single telephoto lens was trained on a two-dimensional roundel target lamped to the bridge handrail. The rounded centre was located and predicatively tracked using a transputer array (Stephen et al. 1993). The system has also been used for monitoring of the Second Severn Crossing (Macdonald et al. 1997) and for applications in a shaking table testing. The system has been patented and marketed, but in the bridge systems had resolution of 0.5mm at 200m range and real-time processing rates up to 12.5Hz.

3.3.2.3 Optical marker tracking

In the biomechanics and entertainment communities several technologies have been developed for tracking movement of the human body (Richards 1999). The same technology has also been used in a wider range of applications and there is limited potential to use the technology for tracking motion of small structures in controlled conditions. Sample rate is up to 200Hz and accuracy depends on the field of view; at 3m it is 0.05mm in transverse direction and 0.3mm in the range axis.

Figure 3-3. Optometer targets at Humber Bridge.

3.3.2.4 GPS

The Global Positioning System (GPS) has begun to see applications for tracking behavior of a number of structures, typically tall buildings and long span bridges. The promise of GPS is that 'absolute' position of the GPS antenna can be determined, to an accuracy of a few mm at sample rates as high as 20Hz. GPS has been used for quasi-static measurements in geotechnical applications such as landslide monitoring (Gili et al. 2007) and on dams (Rutledge and Meyerholtz 2005) while dynamic range measurements have been made on tall buildings (Brownjohn et al. 2005; Celebi 2000; Celebi and Sanli 2002; Kijewski-Correa et al. 2006; Q. S. Li and Wu 2007) and a TV tower (Tamura et al. 2002). Although GPS is a relatively expensive sensing technology and many monitoring programs can suffice with relative response data only, there are applications that warrant GPS use. For example, total wind effects, including the mean and background components not captured by accelerometers, may be of interest (Kijewski-Correa and Kochly 2007).

For civil structures, GPS is usually used in a differential mode called real time kinematic (RTK) where positional errors for a receiver at a single known stable (i.e., stationary) base station are transmitted to the moving receiver or rover. Software is used to obtain a positional fix of the rover that is affected only by the very local conditions and small differences in signal transmission from satellites to the two different receivers.

There are several difficulties with using GPS (Meng et al. 2006; Mickitopoulou et al. 2006), including the various forms of noise and errors that can be introduced such as satellite visibility, multi-path, cycle slip and differential atmospheric effects and data fusion issues. Multipath remains one of the most significant error sources for this application. Quad-constellation choke ring antennas show some promise to mitigate this problem and many commercial receivers incorporate multipath removal algorithms, though removal in post-processing is often still required (Kijewski-Correa and Kochly 2007). In addition, despite the advances in receiver and antenna hardware, as well as software to improve the correction of atmospheric delays and other distortions, GPS remains fundamentally constrained by the optimality of the satellite constellations. Such dilution of precision (DOP) errors can be considerable, particularly in urban zones where neighboring buildings obstruct viable satellites. In this regard, the expansion of satellite services through international efforts like GNSS will increase the density of satellites overhead and improve DOP errors.

3.3.2.5 Surveying and total station

Standard surveying techniques have long been used for tracking structures (Moore 1973). For single point measurements GPS may replace theodolites, but 'total stations' are viable for automated optical surveys, comprising a surveying theodolite and electronic distance measurement (EDM) device operated automatically to line up with an array of targets in a slow sequence. The EDM component is essentially a laser and may operate via 'time of flight' for long range or phase shift for short range measurements. Applications so far have been mainly in geotechnics, but a there are a number of examples of their use for bridge measurement (List et al. 2006; Psimoulis

40   STRUCTURAL IDENTIFICATION OF CONSTRUCTED SYSTEMS

and Stiros 2007). Sample rates for such systems are slower than GPS and accuracy better e.g., as good as 1 arc second and 1mm ±1ppm (range).

Figure 3-4. Radar system and 183m chimney

3.3.2.6 Microwave interferometry

The possibility of simultaneously collecting synchronized displacement signals from distributed locations is realized in a microwave-based system (Bernardini et al. 2007) that works on the same principle as radar. Figure 3-4 shows such a system in use to measure a 183m chimney (Brownjohn et al. 2009). Thirty six points on the structure were tracked simultaneously with a resolution approaching 0.01mm at a sample rate of 50Hz (higher speeds are possible), enabling a partial mode shape to be recovered. The system has also been used successfully for operational modal analysis of a large concrete bridge (Gentile and Bernardini 2008). Such a system is ideal where attachment of traditional sensors is impossible, although performance and ability to track parts of structure depends on the radar reflectivity of the structure and its surroundings.

STRUCTURAL IDENTIFICATION OF CONSTRUCTED SYSTEMS   41

3.3.2.7   Pneumatic systems

Deck profile or level monitoring systems have been installed in a number of bridges. These devices use pneumatic/hydraulic sensors to provide values of height relative to a datum. A comparison of the performance of these devices with GPS at Tsing Ma Bridge (Wong et al. 2001) shows comparable accuracy. A similar system is installed at Tamar Bridge in the UK, and comprises eight sensing locations. These sensors are part of a 70-sensor system installed in 2000 and sampling at 1Hz to track the performance of a bridge upgrade, and are now integrated with a dynamic response monitoring system operated by University of Sheffield (List et al. 2006).

3.3.2.8   Contacting displacement measurements

Relative motion between a structure and a fixed reference at very close range or between parts of structure e.g., across expansion joints can be tracked using a range of instruments including the LVDT (linear variable differential transformer) and the pull-wire extensometer. These are traditional and well understood instruments (Beckwith et al. 1995).

3.3.2.9   Derivation of displacement from acceleration, velocity, strain or rotation signals

Without exception, the technologies previously described require reference stations or fixed positions for measurements of relative displacement. This presents several fundamental limitations. For example there may be no convenient reference position (for laser or total station measurements at long range), especially in built up urban areas. Even when available, atmospheric conditions may degrade the system performance. An alternative solution that dispenses with the need for a stable reference and provides absolute displacements is to use high-precision accelerometers or seismometers, since their signals are derivatives of displacement, which can in principle be recovered by the inverse operation of integration. The significant problem in this case is that numerical integration of the digitized signals results in amplification of the inevitable signal noise at low frequencies ω, in proportion to $1/\omega$ for velocity signals and $1/\omega^2$ for (the more common) acceleration signals.

An exercise conducted with acceleration signals recorded on a tall building (Brownjohn and Pan 2008) during a distant 'great earthquake' showed that components of absolute displacement with frequencies above 0.02Hz could be reliably recovered by integration of acceleration from high-grade servo-accelerometers with stabilized signal conditioning. Figure 3-5 shows integrated signals from the 'boxing day' earthquake recorded at Republic Plaza (Brownjohn and Pan 2008). The rigid body motion of the whole structure is clearly visible from the figure since the movement of the base and level 65 are almost same, a feature not found in the GPS signal.

3.3.3   Velocity

Velocity measurements are a common feature of seismic studies since seismometers usually measure velocity via the current generated in a relative velocity between a

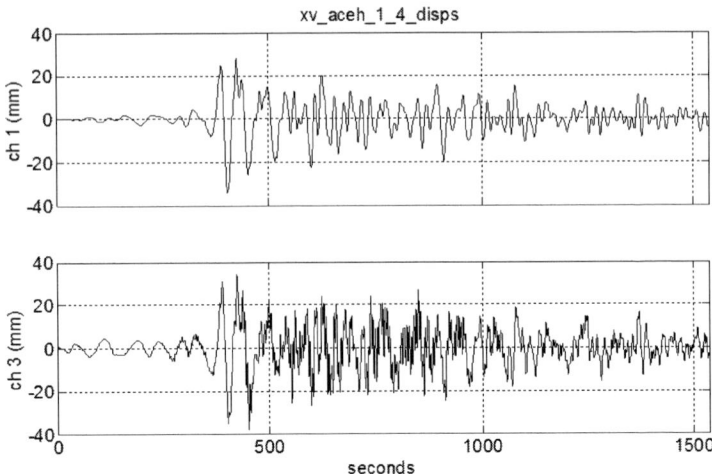

Figure 3-5. Motion of Republic Plaza during Boxing Day earthquake 2004: ch1 corresponds to first basement level, ch3 corresponds to (roof) level 65

coil and a magnet (Usher et al. 1979). Laser Doppler Vibrometers (LDVM) whose applications are described in 3.3.2.1 measure velocity via Doppler shifting of light frequencies. The benefit of their use is that they are non-contacting and can operate at both short and long ranges, but they are generally expensive and can only measure at a single point at a time.

3.3.4 Strain

As for accelerometers there is a similarly large range of strain gauge types, including those based on fiber optic technology such as fiber Bragg grating. Electrical resistance strain gauges are a cheap but often noisy technology, while vibrating wire (Yu and Gupta 2005) strain gauges (with low sample rates) are popular because of reliability and repeatability. Without knowing a baseline value, only differential strain can be measured.

3.3.5 Stress, force

Direct stress measurement instruments that are not simple load cells or variants of strain gauges, are relatively rare and as for strain, only relative values are likely to be available. Vibrating wire stress cells are apparently mainly used for measurements in tunnel linings and have also been used in concrete box-girder bridges (Brownjohn and Moyo 2001). A form of stress cell using elasto-magnetic effects is used to monitoring cable forces e.g., for post-tensioning tendons and stays, main cables and hangers of suspended span bridges (Sumitro et al. 2005). Direct measurements of ice force were used in the Confederation Bridge study (Cheung et al. 1997), for calibrating loading models for offshore installations.

STRUCTURAL IDENTIFICATION OF CONSTRUCTED SYSTEMS 43

Figure 3-6. Anemometer types: sonic, pitot tube/vane, 'windmill'

3.3.6 Pressure

Pressure measurement technology directly parallels force measurement technology using for example vibrating wire sensors (for static measurements) and/or strain gauges (for dynamic measurements) and is well described in standard texts (Doebelin 1990). High-speed pressure measurement on surfaces via pressure taps (Marighetti et al. 2000) is a standard technology in wind tunnel testing but measurements of wind pressure are relatively rare at full-scale, e.g., studies of Great Belt Bridge vortex-shedding response (Frandsen 2001). Dynamic water pressure measurements are needed to study fluid-structure interactions, particularly for dams (Daniell 1994) but static pressure measurements are more common, e.g., in piezometers for water level measurement, as well as for manometer-style devices used for example in structure level sensing.

3.3.7 Temperature

Temperature sensors are usually installed in other instruments, e.g., vibrating wire and fiber optic strain gauges to compensate for thermal effects on instrument performance, otherwise vibrating wire devices, thermocouples and thermistors (Brownjohn and Moyo 2001) and even fiber optic sensors, are used.

3.3.8 Wind

Various forms of anemometer are widely used in full-scale tests of structures, including cup and vane, windmill, propeller, sonic anemometers and forms with clear aerospace origin (Figure 3-6). Other forms (not used for full-scale structural measurements) include hot wire and laser Doppler anemometers (for wind tunnel) and Doppler sonar for meteorology. Cup-and-vane devices are the conventional standard, measuring horizontal component of wind speed and compass bearing. Measurement of all three components of wind requires devices with propellers along three axes or sonic anemometers. Technical factors affecting choice of anemometer include number of components resolved and frequency response, sometimes expressed in terms of a length constant (Brook 1977; Wood 1982) that refers to the 'length' of a gust that must pass an anemometer for it to respond). Practical factors include cost, use of moving parts, susceptibility to electromagnetic interference and form of data output. Problems in using anemometers have been reported by a few researchers e.g., Brownjohn and Pan (2008), mostly in the higher performance types such as the sonic, so reliability should be the main consideration for long term monitoring. With reliability in mind, many bridge monitoring systems use multiple types (S. Li and Wu 2008), with the direction/speed or windmill sensor shown in Figure 3-6 (right) proving particularly reliable.

3.3.9 Mechanical impedance

The electro-mechanical impedance (EMI) technique (Park and Inman 2007) is relatively new entrant in the field of structural health monitoring (SHM), with its origin dating back only to the mid 1990s. Thin patches of the ceramic lead zirconate titanate (PZT), surface-bonded on the host structure, play the key role as 'impedance transducers' in this technique. They act as collocated actuators and sensors and employ ultrasonic vibrations (typically in 30-400 kHz range) to obtain characteristic admittance 'signatures' of the structure. The sensitivity of the PZT patches is high enough to capture structural damage at the incipient stage, well before it acquires detectable macroscopic dimensions.

This first demonstration of the EMI technique on any prototype structure was on a two-span reinforced concrete (RC) bridge, instrumented with several 10x10x0.2mm PZT patches (Soh et al. 2000). The study showed how the patches 'identified' the structure as a parallel spring(k)-damper(c) combination for three load cycles with increasing levels of damage: reduction in the stiffness and increase in the damping is well-known phenomenon associated with crack development in concrete. The main limitation of the EMI technique is the localized zone of influence of the PZT patch as an impedance transducer, in addition, the impedance analyzers used with PZT patches have been cumbersome and expensive, although recently an inexpensive impedance measurement chip.

3.3.10 Corrosion

A major concern for infrastructure operators, particularly for bridges is monitoring corrosion and assessing its effect on the condition of structures. As well as effects on exposed steel structural members, corrosion is a major problem with bridge stay cables and tendons. Significant corrosion has been discovered in the main cables of several long span suspension bridges, including Severn and Forth Road bridges in the UK. Existing corrosion measurement techniques include electrochemical techniques, electrical resistance (ER) probes and measurement of chloride concentration in concrete, as well as destructive (coring) techniques. Some reviews of the technology for detecting corrosion are available for reinforced concrete structures (Broomfield et al. 2002; Hammersley and Dill 1998). Changes in the electrical resistance (ER) of hybrid carbon fiber reinforced polymer (HCFRP) composite sensors have been found suitable to monitor the corrosion of prestressed concrete (PC) tendons, enabling distributed corrosion monitoring for PC structures using multi-electrodes.

For suspension cables and tendons, acoustic emission is a proven technology for detecting wire breaks (Fricker and Vogel 2007), indeed a system installed on the

Forth Road Bridge has been reporting strands snapping at the rate of 'one a month'. The same magneto-elastic sensor technology used to measure cable stresses (Sumitro et al. 2005) has also found applications for monitoring corrosion in steel structures (Singh et al. 2005).

3.3.11 Fiber optic sensors for civil infrastructure

Fiber optic sensors are also commonly used for structural health monitoring. These sensors have certain advantageous characteristics especially because of their insensitivity to external perturbations and electromagnetic interference. There are four main types of FOS for structural and geotechnical measurements (Glisic and Inaudi 2007; Udd 1995): Multiplexed sensors (e.g., fiber Bragg grating or FBG), Long-base sensors (e.g., SOFO) Distributed sensors (Brillouin or Raman), and Point sensors (Fabry-Perot).

In almost all FOS applications, the optical fiber is a thin glass fiber that is protected mechanically by a polymer coating and further protected by a multi-layer cable structure designed to protect the fiber from the installation environment. Since glass is inert and resistant to almost all chemicals, even at extreme temperatures, it is ideal for use in harsh environments, e.g., geotechnical applications. Since the light confined into the core of the optical fibers does not interact with any surrounding electromagnetic field, FOS are immune to any electromagnetic (EM) interferences and are intrinsically safe, making them particularly suitable for monitoring in petrochemical and space applications.

3.3.11.1 Fiber Bragg Grating (FBG) Sensors

A comprehensive discussion of the operation of FBG sensors for applications in structural health monitoring has been published (Todd et al. 2007). Bragg gratings are periodic alterations of the density of glass in the core of an optical fiber, produced by exposing the fiber to intense ultraviolet light. The produced gratings typically have a length of about 10 mm. Light at the wavelength corresponding to the grating period will be reflected while all other wavelengths will pass through the grating undisturbed. The grating period (length) changes with temperature and strain so both parameters can be measured through the spectrum of the reflected light, with accuracy of the order of 1 $\mu\varepsilon$ and 0.1 °C. The main benefit with FBG is their multiplexing potential, with several gratings in the same fiber at different locations and tuned to reflect different wavelengths. FBGs can be used as replacements for conventional strain gages, e.g., by gluing on metals and other smooth surfaces. With adequate protection they can also be used to measure strains in concrete over gauge lengths of around 100 mm. Field applications of FBGs are still relatively rare, e.g., Gebremichael et al. (2005), partly due to their fragility and partly because low-cost of ruggedized field-portable loggers are only recently becoming available.

3.3.11.2 SOFO Interferometric Sensors

SOFO interferometric sensors are long-base sensors, with a measurement base that ranges from 200mm to 10m or more. The SOFO system uses low-coherence interferometry to measure the length difference between two optical fibers installed on the structure to be monitored (Figure 3-7), by embedding in concrete or surface mounting. The measurement fiber is pre-tensioned and mechanically coupled to the structure at two anchorage points in order to follow its deformations, while the reference fiber (in the same pipe) is free and acts as temperature reference. The sensors have excellent long-term stability and accuracy of ±2 μm irrespective of the measurement base.

3.3.11.3 Fabry-Pérot Interferometric Sensors

Fabry-Pérot Interferometric FOS have a single measurement point at the end of the fiber. An extrinsic Fabry-Pérot Interferometer (EFPI) consist of a capillary glass tube containing two partially mirrored optical fibers facing each other, but leaving an air cavity of a few microns between them. When light is coupled into one of the fibers, a back-reflected interference signal is obtained from the two mirrors. This interference can be demodulated to reconstruct the changes in the fiber spacing. Since the two fibers are attached to the capillary tube near its two extremities (with a typical spacing of 10 mm), the gap change will correspond to the average strain variation between the two attachment points. Many sensors based on this principle are currently available for geotechnical monitoring, including piezometers, strain gauges, temperature sensors, pressure sensors and displacement sensors.

Figure 3-7. SOFO sensor installed on a rebar. The plastic pipe contains the coupled measurement fiber and a free un-coupled reference fiber. Metallic anchors at both ends of the white plastic pipe define the gauge length (Source: Daniele Inaudi, reproduced with permission).

3.3.11.4 Distributed Brillouin Scattering and Distributed Raman Scattering Sensors

Distributed FOS measure physical parameters, in particular strain and temperature, along their whole length, allowing the measurements of thousands of points from a single readout unit. If an intense light at a known wavelength is shone into a fiber, a very small amount of it is scattered back from every location along the fiber itself. Besides the original wavelength (called the Rayleigh component), the scattered light contains components at wavelengths that are higher and lower than the original signal (called the Raman and Brillouin components). These shifted components contain information on the local properties of the fiber, in particular its strain and temperature. Systems based on Raman scattering typically exhibit temperature accuracy of the order of ± 0.1°C and a spatial resolution of 1m over a measurement range up to 8 km. The best Brillouin scattering systems (Karashima 1990) offer a temperature accuracy of ± 0.1°C, a strain accuracy of ±20 microstrain and a measurement range of 30 km, with a spatial resolution of 1 m.

3.3.12 Hybrid carbon fiber reinforced polymer (HCFRP) sensors

A novel type of hybrid carbon fiber-reinforced polymer (HCFRP) sensing techniques for the SHM civil infrastructure has been developed (Wu and Yang 2005). Characterized by low cost, long gauge length, long-term and distributed sensing as well as long-term durability; HCFRP sensing technique is based on the piezoresistivity and electrical conduction of carbon fibers. The electrical resistance (ER) measurement of the full length of the HCFRP sensors provides for global monitoring for structures. By installing multi-electrodes to the HCFRP sensor, distributed sensing can be achieved by measuring the ERs between every two electrodes, while change in ER of the whole length of the sensor clearly indicates damage (Yang and Wu 2006).

3.4 Data Transmission

3.4.1 Wired and fiber optic connections

Data transmission using wired technology is a key component of the majority of monitoring systems. For permanent installation, cable with relevant ratings such as low smoke zero halogen (LSZH) and high ingress protection (IP) must also be structurally robust (particularly exposed on wind/rain-blown faces of structures) and have the required electrical characteristics of low resistance and shielding of conductors. High quality cable will minimize problems with electromagnetic interference and cross-talk. Measurement of cables (such as due to wind) can affect low voltage signals such as resistance based strain gages. Lightning protection may be a second concern for exposed wiring, sensors and housings, requiring careful arrangements for grounding (avoiding earth loops) and surge-suppression to avoid high voltage differences and flashover adjacent to sensors and signal conditioning.

Wired systems for temporary, time limited studies, such as vibration surveys require robust and foolproof but quick-fit connectors, while permanent installations require connectors with high IP ratings, e.g., the Mil-Standard connectors used with seismic monitors. Where ICP/IEPE accelerometers are used, standard microdot cables are kept to a minimum as these can be noisy especially when mishandled.

In cases where multiple loggers need to be networked using local area network connections, or where modems and logger are separated, fiber optic links are required for distances over 100m. Synchronization of signals from loggers is an acute problem for dynamic testing, most usually dealt with using hard-wired analog connections. For merging digitized data streams from separate loggers, wireless sensors and 'internet accelerometers' synchronization is a prime concern, that may limit the effective bandwidth of the sensing solution.

3.4.2 Wireless Sensors for Structural Monitoring

In response to the high costs associated with tethered structural monitoring systems, researchers in the civil engineering community have proposed the use of wireless communication for the transfer of data between sensors and a data repository in future structural monitoring systems. Such wireless monitoring systems are assembled from low-cost wireless sensors that collocate sensing, communication and computing in a single device (Lynch 2002; Spencer et al. 2004; Straser and Kiremidjian 1998). For interested readers, Lynch and Loh (2006) provide a more complete review of the applications of wireless sensors for structural health monitoring application up to 2006.

An alternate simple form of wireless monitoring is to use synchronized autonomous recorders, with post-processing to splice data records together. Accurate timing is provided by GPS antennae, but delays of the order of milli-seconds cannot be avoiding, limiting such a system to applications on structures with natural frequencies well below 10 Hz. Also such a system is not real-time operation and it is not possible to exercise quality control and check sensor operation during the measurements.

3.5 Data Acquisition and Management

Data acquisition is the procedure for converting analog or digital signals, transmitted from sensors by wired or wireless links, to digital data. These data may be permanently stored locally (the default choice for most acquisition systems, on a computer disk drive), alternatively the data may be processed locally to a reduced quantity of higher level data/information. This applies to both short- term field investigations (e.g., modal surveys) and medium or long-term monitoring, but system architecture for long term monitoring usually caters for a broader range of signal types compared to dynamically varying acceleration-equivalent voltages acquired in modal tests, hence we distinguish between the two applications here.

3.5.1 Data acquisition for short/long term structural monitoring

A wide range of measurements are covered here, ranging from slowly sampled (static) signals already in digital form to conventional analog (voltage) signals at varying dynamic sample rates. Capturing diurnal variation of static response parameters (e.g., temperature) can be accomplished by sampling as slowly as once per hour.

For dynamic signals, sample rates would depend on the structure size and frequency range of the loading. For global response of long span bridges (>500m) and tall buildings (>200m) 10 Hz bandwidth is more than adequate, requiring sample rates approximately 2.5 times larger to provide room for anti-alias filtering. For short span highway bridges and pedestrian structures, bandwidth up to 40 Hz will suffice. Seismometers typically default to 100Hz bandwidth.

In this category come general purpose logger systems with pure data acquisition function which can be configured to read a wide range of sensor types (analog voltage, vibrating wire gauges, thermocouples, strain gauge bridges, etc.) traditionally at slow sample rates, but sometimes with high speed acquisition capabilities. Seismometers and logger/interfaces for fiber optics, GPS and other exotic signal types come in this category as their functionality is limited. Such loggers may be networked (e.g., via 'multi-drop' systems such as IEEE 1451.3 or Ethernet) and interrogated directly (via modem or data card) or controlled by PC (Brownjohn and Moyo 2001; Moyo et al. 2004).

Reliability and longevity of these systems requires robust hardware: protection from dust, overheating and moisture, protection against theft and vandalism, rugged PCs, redundant data storage (e.g., RAID (redundant array of independent disks) disk drives). Increasingly, upgrade paths will be a concern for monitoring systems expected to last as long as a decade. Reliable and clean power supplies with UPS (uninterruptible power supplies) protection must be provided as well as communication via high speed broadband links.

3.5.2 Data acquisition for modal surveys

Data acquisition systems range from proprietary boxes with comprehensive built-in software for both acquisition and signal analysis to component-based systems programmable by skilled users, and all but the simplest provide capability for signal generation, e.g., for shaker control. While systems traditionally used for laboratory-based testing may work in field conditions of mains power and shelter, fully field-portable systems with maximum flexibility and mobility have additional requirements including light weight, weather-resistance and capability for battery operation, in which case a smaller channel count is likely to be optimal. Data acquisition systems for modal testing are grouped under four categories:

- Systems primarily aimed at automotive/aerospace users work directly with sophisticated embedded signal/modal analysis software (described in a later chapter) to generate modal data (frequencies, mode shapes, damping ratios, modal constants).

- Multi-channel spectrum analyzers that record signals and convert to frequency domain, generating cross-spectral density (CSD) matrices that form the basis for modal analysis by separate software. Original time series may be discarded once CSD matrices have been generated.

- Networked or stand-alone seismometers that measure and record ground motions created by earthquakes and other sources.

- Component-based modular systems offer a highly flexible and cost effective solution for field data acquisition, mainly for permanent system but also for low-cost modal tests. The trade off is between flexibility and skill level of the user/programmer.

For proprietary systems, the user's choice of analog to digital conversion hardware parameters is limited, traditionally to higher capabilities in sample rates, precision (high dynamic range) and sharp filtering (which may be provided by over-sampling, digital filtering and decimation). 24-bit analog to digital conversion, with ratio of largest signal to single bit level set at $\pm 2^{23}$, allows for signal dynamic range exceeding 130dB and is becoming the default choice for data acquisition software. Such resolution is more than adequate to cover the full dynamic range of the best seismometers so that selecting signal ranges ceases to be a concern for such systems.

For the modular systems, while 24-bit technology is becoming the default, choices still need to be made about maximum sample rate per channel, buffering for simultaneous sample and hold (SSH, where there is no delay or 'skew' as the single ADC multiplexes between adjacent channels) and type of anti-aliasing filter. Control of these systems is almost universally by micro-computer: with PC cards embedded in desktop or semi-portable PCs or communicating with hardware platforms via interfaces such as IE488. More portable solutions now use either embedded PCs/hardware interface via LAN connections, USB or firewire.

3.5.3 Data Storage, file management, archiving.

This is the final link in the chain before mathematical treatment of data, and also applies as much to raw data as to parameters condensed from raw data via embedded (local) processing. The most efficient data storage format is most likely to be binary files generated directly by acquisition software. Rather than saving directly to a database, an efficient procedure is individual files saved at convenient intervals (e.g., 10 minutes, one hour or one day) with data stamped file names. File sizes depend on channel count and sample rate; while disk storage has ceased to be a significant constraint except when dealing with video recordings, using sensible sample rates will speed up internet transfer and processing.

While raw data will typically remain on the local storage system, they still need to be accessed and viewed to investigate key structural events (e.g., response to storm or earthquake). To facilitate this, summaries may be generated by automated local processing, saving statistical values (e.g., mean, variance) of data channels in files that may be small enough to email from the remote PC. For time-limited measurements (short measurement campaigns or complete monitoring exercises) raw test data have high value, hence good practice requires careful and logically organized data archiving along with all records of the testing including specifications, plans, method statements, notes on sensor configurations and calibrations, photographs, videos and drawings. Meticulous management allows for handover to unfamiliar engineers for operation and re-analysis. Where processed results are saved, the version of software used to generate them is also saved and for raw binary data a version of reader software and/or details of the file structure safeguard future accessibility.

3.6 Use of Non-Destructive Evaluation for Structural Identification

Non-destructive evaluation (NDE) applications have greatly progressed over the last decade especially for bridge type structure's inspections. The dominant practice by state DOTs in evaluation of bridge decks is by visual inspection and use of simple nondestructive methods like chain drag and hammer sounding. Modern nondestructive evaluation of concrete and concrete bridge decks has its origins in geophysics. A number of techniques introduced exploit various physical phenomena (acoustic or seismic, electric, electromagnetic, thermal, etc.) to detect and characterize specific deterioration processes or defects, as summarized in Table 3-1.

Table 3-1. NDE techniques and their application to bridge deck deterioration/defect detection and characterization

	NDT Method	Defect/deterioration applications	Other applications
Electro-Magnetic	Ground Penetrating Radar (GPR)	Deterioration of concrete induced by corrosion, salt and acid actions, water penetration. Indirect delamination detection.	Thickness of the deck, concrete cover, rebar location.
	Electrical Conductivity	Damage to rebars and tendons	
	Spectral Induced Polarization (SIP)	Detection of voids and presence of moisture	
Acoustic/ Seismic	Impact Echo (IE)	Corrosion induced deck delamination detection and characterization.	Thickness of the deck. Investigation of crack
	Ultrasonic-echo (UPE)	Detection of voids and other anomalies.	Localization of rebars and tendons
	Ultrasonic Surface Waves (USW)	Measurement of degradation of mechanical properties (modulus, strength)	
	Ultrasonic Transmission (UPV)	Measurement of degradation of mechanical properties, detection of voids and cracks	Reinforcement and tendon ducts detection
Chemical/ Potential	Potential mapping	Corrosion of reinforcement	
	Laser Induced Breakdown Spectroscopy (LISB)	Near surface analysis of ingress of chemicals	
Thermal	Infrared (IR) Thermography	Detection of debonding of overlays, delamination, presence of moisture and near surface voids.	

In general, all the techniques utilize an approach where the objective is to learn about the characteristics of the medium from the response of the medium to the applied excitation. Some mechanisms primarily affect the reinforcement and some theconcrete itself, but all degradation mechanisms lead to a less resistant structure and, thus, promote other deterioration mechanisms. Steel corrosion is probably the most commonly encountered deterioration mechanisms of reinforced concrete, and generally deterioration of the highest concern. Main deterioration of concrete itself caused by chemical mechanisms are alkali aggregate reactions, acid attacks and sulfate reactions. The following sections identify main deterioration mechanisms in concrete bridge decks as an example. NDE applications to other type of structures and structural elements are available in the literature. Detailed discussions of these methods are beyond the scope of this report.

3.7 Closing Remarks

This chapter has attempted to provide an overview of the experimental technology for structural identification of (civil) constructed systems. Since technology continually evolves, particularly in relation to data acquisition and storage hardware and software, this is a snapshot of a moving target. However, there are valuable lessons and principles to be extracted from past exercises, with history repeating itself. Regrettably there are still not nearly enough full-scale experimental studies, and many of those are not well publicized or provide limited details. We have tried to report on as many as possible, and it is always important to exchange and publicize the experiences of researchers, practicing engineers and owners who take the trouble to investigate operational structures at full-scale.

3.8 References

Ahola, R., and Tervaskanto, M. (1991). "A position-sensing laser system for analyzing dynamic deflection in large constructions." *Field Measurements in Geotechnics,* Oslo, Norway.

Anderson, D., and Mills, B. (1971). "Multi-point excitation techniques." *Environmental Engineering,* 51, 12-16.

Avitabile, P. (1998). "Modal space - back to basics: Is there any difference between a roving hammer and roving accelerometer test?" *Experimental Techniques,* 22(5), 9-10.

Bakht, B., and Jaeger, J. G. (1990). "Bridge testing: A surprise every time." *ASCE Journal of Structural Engineering,* 116(5), 1370.

Beckwith, T. G., Marangoni, R. D., and Lienhard, J. H. (1995). *Mechanical measurements,* Addison-Wesley Publishing, Reading, MA.

Bell, E. S., Sipple, J., and Yost, J. (2008). "Long-term thermal performance of a CFRP-reinforced bridge deck." *Structures Congress,* ASCE, Vancouver, Canada.

Bernardini, G., De Pasquale, G., Bicci, A., Marra, M., Coppi, F., and Ricci, P. (2007). "Microwave interferometer for ambient vibration measurement: I. principles of the radar technique and laboratory tests." *EVACES '07: International Conference on Experimental Vibration Analysis*, Porto, Portugal.

Bougard, A. J., and Ellis, B. R. (2000). "Laser measurement of building vibration and displacement." *Shock and Vibration*, 7, 287-298.

Brook, R. R. (1977). *Boundary-Layer Meteorology*, 11(1), 33-37.

Broomfield, J. P., Davies, K., and Hladky, K. (2002). "The use of corrosion monitoring in new and existing reinforced concrete structures." *Cement and Concrete Composite*, 24, 27-34.

Brownjohn, J. M. W. (2007a). "Noise characteristics of sensors for extreme low level vibration measurements." *25th International Modal Analysis Conference*, Society of Experimental mechanics, Orlando, FL.

Brownjohn, J. M. W. (2007b). "Structural health monitoring of civil infrastructure." *Philosophical Transactions of the Royal Society of London, Series A, Mathematical and Physical Sciences*, 365, 589-622.

Brownjohn, J. M. W., Carden, E. P., Goddard, C. R., Oudin, G., and Koo, K. (2009). "Real-time performance tracking on a 183m concrete chimney and tuned mass damper system." *IOMAC 2009*, Porto Novo, Italy.

Brownjohn, J. M. W., Dumanoglu, A. A., and Severn, R. T. (1992). "Ambient vibration survey of the Fatih Sultan Mehmet (second bosporus) suspension bridge." *Earthquake Engineering and Structural Dynamics*, 21, 907-924.

Brownjohn, J. M. W., Fok, P., Roche, M., and Moyo, P. (2004). "Long span steel pedestrian bridge at Singapore Changi airport - part 1: Prediction of vibration serviceability." *Structural Engineer*, 82(16), 21-27.

Brownjohn, J. M. W., Fok, P., Roche, M., and Omenzetter, P. (2004). "Long span steel pedestrian bridge at Singapore Changi airport - part 2: Crowd loading tests and vibration mitigation measures." *Structural Engineer*, 82(16), 28-34.

Brownjohn, J. M. W., and Moyo, P. (2001). "Monitoring of Malaysia-Singapore second link during construction." *2nd International Conference on Experimental Mechanics*, Singapore.

Brownjohn, J. M. W., Moyo, P., Omenzetter, P., and Lu, Y. (2003). "Assessment of highway bridge upgrading by dynamic testing and finite element model updating." *ASCE Journal of Bridge Engineering*, 8(3), 162-172.

Brownjohn, J. M. W., and Pan, T. C. (2008). "Identifying loading and response mechanisms from ten years of performance monitoring of a tall building." *ASCE Journal of Performance of Construction Facilities,* 22(1), 25-34.

Brownjohn, J. M. W., Pavic, A., and Omenzetter, P. (2004). "A spectral density approach for modeling continuous vertical forces on pedestrian structures due to walking." *Canadian Journal of Civil Engineering,* 31(1), 65-77.

Brownjohn, J. M. W., Stringer, M., Tan, G. H., Poh, Y. K., Ge, L., and Pan, T. C. (2005). "Experience with RTK-GPS system for monitoring wind and seismic effects on a tall building." *SHMII-2: Structural Health Monitoring and Intelligent Infrastructures,* Shenzen, China.

Brownjohn, J. M. W., and Tao, N. F. (2005). "Vibration excitation and control of a pedestrian walkway by individuals and crowds." *Journal of Shock and Vibration,* 12(5), 333-347.

Buckland, P. G., Hooley, R., Morgenstem, B. P., Rainer, J. H., and Van, S. (1979). "Suspension bridge vibrations: Computed and measured." *ASCE Journal of Structural Engineering,* 105(5), 859-874.

Caetano, E., Cunha, A., and Moutinho, C. (2007). "Implementation of passive devices for vibration control at Coimbra footbridge." *Experimental Vibration Analysis of Civil Engineering Structures,* FEUP,

Caetano, E., Silva, S., and Bateira, J. (2007). "Application of a vision system to the monitoring of cable structures." *7th International Symposium on Cable Dynamics,* Vienna, Austria.

Calcada, A., Cunha, A., and Delgado, R. (2005). "Analysis of traffic-induced vibrations in a cable-stayed bridge. part 1: Experimental assessment." *ASCE Journal of Bridge Engineering,* 10(4), 370-385.

Carne, T. G., and Stasiunas, E. C. (2006). "Lessons learned in modal testing - part 3: Transient excitation for modal testing, more than just hammer impacts." *Experimental Techniques,* 30(3), 69-79.

Catbas, F. N., Ciloglu, S. K., Hasancebi, O., Grimmelsman, K. A., and Aktan, A. E. (2007). "Limitations in structural identification of large constructed structures." *Journal of Structural Engineering,* 133(8), 1051-1066.

Cebon, D. (1993). *Interaction between heavy vehicles and roads,* SP-0951, SAE International, Warrendale, PA.

Celebi, M. (2000). "GPS in dynamic monitoring of long-period structures." *Soil Dynamics and Earthquake Engineering,* 20, 477-483.

Celebi, M., and Sanli, A. (2002). "GPS in pioneering dynamic monitoring of long-period structures." *Earthquake Spectra*, 18(1), 47-61.

Cheung, M. S., Tadros, G. S., Brown, J., Dilger, W. H., Ghali, A., and Lau, D. T. (1997). "Field monitoring and research on performance of the confederation bridge." *Canadian Journal of Civil Engineering*, 25(951), 962.

Cunha, A., Caetano, E., and Magalhaes, F. (2006). *From input-output to output-only modal identification of civil engineering structures*, SAMCO Association, Europe.

Cunha, A., Caetano, E., and Magalhaes, F. (2007). "Output-only dynamic testing of bridges and special structures." *Structural Concrete*, 8(2), 67-85.

Dalgliesh, W. A., and Rainer, J. H. (1978). "Measurements of wind induced displacements and acceleration of a 57-storey building in Toronto, Canada." *3rd Colloquium on Industrial Aerodynamics*, Aachen, Germany.

Daniell, W. E. (1994). "Full-scale dynamic testing and analysis of a reservoir intake tower." *Earthquake Engineering and Structural Dynamics*, 23(11), 1219-1237.

Doebelin, E. O. (1990). *Measurement systems: Application and design*, McGraw-Hill, New York, NY.

Dougill, J. W., Wright, J. R., Parkhouse, J. G., and Harrison, R. E. (2006). "Human structure interaction during rhythmic bobbing." *The Structural Engineer*, 84(22), 32-39.

Dunnicliff, J. (1994). *Geotechnical instrumentation for monitoring field performance*, John Wiley and Sons, Chichester, UK.

Dziuba, P., Grillaud, G., Flamand, O., Sanquier, S., and Tetard, Y. (2001). La Passerelle Solferino: Comportement dynamique (Solferino Bridge: Dynamic behaviour). *Bulletin ouvrages metalliques*, 1, 34-57. c

Esfandiari, A., Bakhtiari-Nejad, F. Rahai, A. and Sanayei, M., "Structural Model Updating Using Frequency Response Function and Quasi-Linear Sensitivity Equation," Journal of Sound and Vibration, doi:10.1016/j.jsv.2009.07.001, 326, September 2009, pp. 557–573.

Esfandiari, A., Bakhtiari-Nejad, F., Sanayei, M., and Rahai, A., "Structural Finite Element Model Updating Using Transfer Function Data," Computers and Structures, doi:10.1016/j.compstruc.2009.09.004, Vol. 88, Issues 1-2, January 2010, pp. 54-64.

Esfandiari, A., Sanayei, M., Bakhtiari-Nejad, F., and Rahai, R., "Finite Element Model Updating using Frequency Response Function of Incomplete Strain Data,"

AIAA Journal, doi:10.2514/1.J050039, 2010. Vol. 48, No. 7, July 2010, pp. 1420-1433.

Ewins, D. J. (1999). "Virtual modal testing." *Asia Pacific Vibration Conference*, Singapore.

Eyre, R., and Tilly, G. P. (1977). "Damping measurements on steel and composite bridges." *Symposium on Dynamic Behavior of Bridges, TRRL Supplementary Report 275*, Transport and Road Research Laboratory, Crowthorne, UK.

Fitzpatrick, A., Dallard, P., le Bourva, S., Low, A., Ridsill Smith, R., and Wilford, M. (2001). *Linking London: The millennium bridge*, Royal Academy of Engineering, London, UK.

Frandsen, J. B. (2001). "Simultaneous pressure and accelerations measured full-scale on the great belt east suspension bridge." *Journal of Wind Engineering and Industrial Aerodynamics*, 89(1), 95-129.

Fricker, F., and Vogel, T. (2007). "Site installation of a continuous acoustic monitoring." *Construction and Building Materials*, 21, 501-510.

Gebremichael, Y. M., Li, W., Meggitt, B. T., Boyle, W. J. O., Grattan, K. T. V., McKinley, B., Boswell, L. F., Aarmes, K. A., Aasen, S. E., Tynes, B., Fonjallaz, Y., and Triantafillou, T. (2005). "A field deployable, multiplexed Bragg grating sensor system used in an extensive highway bridge monitoring evaluation test." *IEEE Sensors Journal*, 5(3), 510-519.

Gentile, C., and Bernardini, G. (2008). "Output-only modal identification of a reinforced concrete bridge from radar-based measurement." *NDT & E International*, 41(7), 544-553.

Gentile, C., and Cabrera, F. (1997). "Dynamic investigation of a repaired cable-stayed bridge." *Earthquake Engineering and Structural Dynamics*, 26(1), 41-59.

Gili, J. A., Corominas, J., and Rius, J. (2007). "Using global positioning system techniques in landslide monitoring." *Engineering Geology*, 55(3), 167-192.

Glanville, M. J., Kwok, K. C. S., and Denoon, R. O. (1996). "Full-scale damping measurements of structures in Australia." *Journal of Wind Engineering and Industrial Aerodynamics*, 59, 349-364.

Glisic, B., and Inaudi, D. (2007). *Fibre optic methods for structural health monitoring*, John Wiley and Sons, New York, NY.

Green, M. F., and Cebon, D. (1994). "Dynamic response of highway bridges to heavy vehicle loads: Theory and experimental validation." *Journal of Sound and Vibration,* 170(1), 51-78.

Hammersley, G. P., and Dill, M. J. (1998). "The long-term monitoring of civil engineering and building structures - developments in techniques for monitoring corrosion in reinforced and post-tensioned concrete." *I MECH E Part I Journal of Systems and Control Engineering,* 212(3), 175-188.

Haritos, N., Hira, A., Mendis, P., Heywood, R., and Giufre, A. (2000). "Load testing to collapse limit state of Barr creek bridge." *Transportation Research Record,* 1696, 92-102.

Higashihara, H., Moriya, T., and Tajima, J. (1987). "Ambient vibration test of an anchorage of south Bisan-Seto suspension bridge." *Earthquake Engineering and Structural Dynamics,* 15, 679-695.

Hudson, D. E. (1964). "Response testing of full-scale structures." *ASCE Journal of Engineering Mechanics,* 90(3), 1-19.

Hudson, D. E. (1970). "Dynamic tests of full-scale structures." *Earthquake engineering,* Prentice-Hall, Inc., Englewood Cliffs, NJ, 127-149.

Hudson, D. E. (1976). "Dynamic tests of full-scale structures." *Dynamic Response of Structures: Testing Methods and System Identification,* University of California, Los Angeles, Los Angeles, CA.

Huston, D. (2011). *Structural sensing, health monitoring and performance evaluation,* Taylor and Francis Group, LLC, Boca Raton, FL.

Hutin, C. (2000). "Modal analysis using appropriate excitation techniques." *Sound and Vibration,* December, 1.

Ibanez, P. (1979). *Review of analytical and experimental techniques for improving structural dynamic problems,* 249, Welding Research Council, Shaker Heights, OH.

Jeary, A. P., and Ellis, B. R. (1981). "Vibration tests of structures at varied amplitudes." *Dynamic response of structures: Experimentation observation prediction and control,* G. Hart, ed., ASCE, New York, 281-294.

Karashima, T. (1990). "Distributed temperature sensing using simulated Brillouin scattering in optical silica fibers." *Optics Letters,* 15, 1038-1040.

Kijewski-Correa, T., Kareem, A., and Kochly, M. (2006). "Experimental verification and full-scale deployment of global positioning systems to monitor the dynamic response of tall buildings." *Journal of Structural Engineering,* 132(8), 1242-1253.

Kijewski-Correa, T., Kilpatrick, J., Kareem, A., Kwon, D. K., Bashor, B., Kochly, M., Young, B. S., Abdelrazaq, A., Galsworthy, J. K., Isyumov, N., Morrish, D., Sinn, R. C., and Baker, W. F. (2006). "Validating wind-induced response of tall buildings: Synopsis of the Chicago full-scale monitoring program." *ASCE Journal of Structural Engineering,* 132(10), 1509-1523.

Kijewski-Correa, T., and Kochly, M. (2007). "Monitoring the wind-induced response of tall buildings: GPS performance and the issue of multipath effects." *Journal of Wind Engineering and Industrial Aerodynamics,* 95(9-11), 1176-1198.

Kramer, C., and De Smet, C. A. M. (1999). "Comparison of ambient and forced vibration testing of civil engineering structures." *17th International Modal Analysis Conference (IMAC-XVII),* Society of Experimental Mechanics, Kissimmee, FL.

Lee, P. K. K., Ho, D., and Chung, H. W. (1987). "Static and dynamic tests of concrete bridge." *ASCE Journal of Structural Engineering,* 113(1), 61-73.

Li, Q. S., and Wu, J. R. (2007). "Time-frequency analysis of typhoon effects on a 79 storey building." *Journal of Wind Engineering and Industrial Aerodynamics,* 95, 1648-1666.

Li, S., and Wu, Z. S. (2008). "Development of distributed long-gage fiber optic sensing system for structural health monitoring." *Structural Health Monitoring,* 6(2), 133-143.

List, D., Cole, R., Wood, T., and Brownjohn, J. M. W. (2006). "Monitoring performance of the Tamar suspension bridge." *3rd International Conference on Bridge Maintenance, Safety and Management,* IABMS, Porto, Portugal.

Littler, J. D., and Ellis, B. R. (2007). "Interim findings from full-scale measurements at Hume point." *Journal of Wind Engineering and Industrial Aerodynamics,* 36, 1181-1190.

Luscher, D. J., Brownjohn, J. M. W., Sohn, H., and Farrar, C. (2001). "Modal parameter extraction of Z24 bridge data." *19th International Modal Analysis Conference,* Society of Experimental Mechanics, Kissimmee, FL.

Lynch, J. P. (2002). *Decentralization of wireless monitoring and control technologies for smart civil structures,* 140, John A. Blume Engineering Center, Stanford, CA.

Lynch, J. P., and Loh, K. J. (2006). "A summary review of wireless sensors and sensor networks for structural health monitoring." *Shock and Vibration Digest,* 38(2), 91-128.

Macdonald, J. H. G., Dagless, E. L., Thomas, B. T., and Taylor, C. (1997). "Dynamic measurements of the Second Severn Crossing." *Proceedings of Institution of Civil Engineers, Transport*, 123(4), 241-248.

Marecos, J., Castanheta, M., and Trigo, J. T. (1969). "Field observations of Tagus River suspension bridge." *ASCE Journal of the Structural Division*, 95(ST4), 555-583.

Marighetti, J., Wittwer, A., De Bortoli, M., Natalini, B., Paluch, M., and Natalini, M. (2000). "Fluctuating and mean pressure measurements on a stadium covering in wind tunnel." *Journal of Wind Engineering and Industrial Aerodynamics*, 84(3), 321-328.

McConnell, K. G. (1995). *Vibration testing theory and practice*, John Wiley and Sons, Inc., New York, NY.

Meng, X., Roberts, G. W., Dodson, A. H., and Brown, C. J. (2006). "GNSS for bridge deformation: Limitations and solutions." *3rd International Conference on Bridge Maintenance, Safety and Management*, International Association for Bridge Maintenance and Safety,

Mickitopoulou, A., Protopsalti, K., and Stiros, S. (2006). "Monitoring dynamic and quasi-static deformations of large flexible engineering structures with GPS: Accuracy, limitations and promises." *Engineering Structures*, 28, 1471-1482.

Miller, R. A., Aktan, A. E., and Sharooz, B. M. (1992). "Nondestructive and destructive testing of a three span skewed R.C. slab bridge." *Conference on Nondestructive Testing of Concrete Elements and Structures*, American Society of Civil Engineers, San Antonio, TX.

Moore, J. F. A. (1973). "The photogrammetric measurement of constructional displacements of a rockfill dam." *Photogrammetric Record*, 7(42), 628-648.

Moyo, P., Brownjohn, J. M. W., and Omenzetter, P. (2004). "Highway bridge live loading assessment and load carrying capacity estimation using a health monitoring system." *Structural Engineering and Mechanics*, 18(5), 609-626.

Myrvoll, F., DiBiagio, E., and Hansvold, C. (1994). "Instrumentation for monitoring the Skarnsundet cable-stayed bridge." *3rd Symposium on Straits Crossings*,

OECD. (1998). *Dynamic interaction between vehicles and infrastructure experiment*, DSTI/DOT/RTR/IR6(981)I, Organization for Economic Cooperation and Development, Paris, France.

Park, G., and Inman, D. J. (2007). "Structural health monitoring using piezoelectric impedance measurements." *Philosophical Transactions of the Royal Society of London, Series A, Mathematical and Physical Sciences*, 365(1851), 373-392.

Partridge, A., and Kenny, T. (2000). "A high performance planar piezoresistive accelerometer." *Journal of Microelectronical Systems,* 9(1), 58-66.

Patron-Solares, A., Cremona, C., Bottineau, C., Leconte, R., and Goepfer, F. (2005). "Study of bell swinging induced vibrations of bell tower of Metz cathedral (France)." *Experimental Vibration Analysis for Civil Engineering Structures,* Bordeaux, France.

Peeters, B., Cornelis, B., Janssens, K., and van der Auweraer, H. (2007). "Removing disturbing harmonics in operational modal analysis." *International Operational Modal Analysis Conference,* Copenhagen, Denmark.

Peeters, B., and Ventura, C. E. (2003). "Comparative study of modal analysis techniques for bridge dynamic characteristics." *Mechanical Systems and Signal Processing,* 17(5), 965-988.

Pernica, G. (1983). "Dynamic live loads at a rock concert." *Canadian Journal of Civil Engineering,* 10(2), 185-191.

Psimoulis, P., and Stiros, S. (2007). "Measurement of deflections and of oscillation frequencies of engineering structures using robotic theodolites (RTS)." *Engineering Structures,* 29, 3312-3324.

Rainer, J. H. (1979). "Dynamic testing of civil engineering structures." *3rd Canadian Conference on Earthquake Engineering,*

Reese, R. T., and Kawahara, W. A. (1993). *Handbook on structural testing,* Fairmont Press, Inc., Bethel, CT.

Reynolds, P., and Pavic, A. (2000). "Impulse hammer versus shaker excitation for the modal testing of building floors." *Experimental Techniques,* 24(3), 39-44.

Reynolds, P., and Pavic, A. (2006). "Vibration performance of a large cantilever grandstand during an international football match." *ASCE Journal of Performance of Construction Facilities,* 20(3), 202-212.

Reynolds, P., Pavic, A., and Carr, J. (2007). "Experimental dynamic analysis of the Kingston communications stadium." *The Structural Engineer,* 85(8), 33-39.

Richards, J. G. (1999). "The measurement of human motion: A comparison of commercially available systems." *Human Movement Science,* 18, 589-602.

Rossi, G., Marsili, R., Gusella, V., and Gioffre, M. (2002). "Comparison between accelerometer and laser vibrometer to measure traffic excited vibrations on bridges." *Shock and Vibration,* 9(1-2), 11-18.

Rubin, S. (1980). "Ambient vibration survey of offshore platform." *ASCE Journal of Engineering Mechanics,* 106(3), 425-441.

Rutledge, D. R., and Meyerholtz, S. Z. (2005). "Performance monitoring of Libby dam with a differential global positioning system." *25th United States Society on Dams (USSD) Annual Meeting,* Salt Lake City, Utah.

Safford, F. B., and Masri, S. F. (1981). "Development and use of a pulse train generator." *Proceedings of the Second Specialty Conference on Dynamic Response of Structures: Experimentation, Observation, Prediction and Control,* Atlanta, GA.

Sanayei, M. and Javdekar C. N., "Sensor Placement for Parameter Estimation of Structures using Fisher Information Matrix," Proceedings of 7th International Conference on Applications of Advanced Technology in Transportation (AATT 2002), Cambridge, MA, August 5-7 2002, pp. 386-393.

Schiff, A. J. (1972). "Identification of large structures using data from ambient and low level excitations." *Annual meeting of the ASME, system identification of vibrating structures: Mathematical models from test data,* W. Pilkey and R. Cohen, eds., American Society of Mechanical Engineers, New York, NY, 87-120.

Severn, R. T., Jeary, A. P., and Ellis, B. R. (1980). "Forced vibration tests and theoretical studies on dams." *Institution of Civil Engineers, Part 2,* 69(3), 605-634.

Singh, V., Wang, M. L., and Lloyd, G. M. (2005). "Measuring and modeling of corrosion in structural steels using magnetoelastic sensors." *Smart Materials and Structures,* 14(3), S24-S31.

Siringoringo, D. M., and Fujino, Y. (2006). "Observed dynamic performance of the Yokohama Bay bridge from system identification using seismic records." *Journal of Structural Control and Health Monitoring,* 13(1), 226-244.

Soh, C. K., Tseng, K. K. H., Bhalla, S., and Gupta, A. (2000). "Performance of smart piezoceramic patches in health monitoring of a RC bridge." *Smart Materials and Structures,* 9(4), 533-542.

Sohn, H., Farrar, C. R., Hemez, F. M., Czarnecki, J. J., Shunk, D. D., Stinemates, D. W., and Nadler, B. R. (2004). *A review of structural health monitoring literature: 1996-2001,* LA-13976-MS, Los Alamos National Laboratory, Los Alamos, NM.

Spencer, B. F., Ruiz-Sandoval, M. E., and Kurata, N. (2004). "Smart sensing technology: Opportunities and challenges." *Journal of Structural Control and Health Monitoring,* 11(4), 349-368.

Spidsoe, N., and Hilmarsen, B. (1983). "Measured dynamic behavior of north sea gravity platforms under extreme environmental conditions." *15th Offshore Technology Conference*, Houston, TX.

Srinivasan, M. G. (1984). *Dynamic testing of as-built civil engineering structures - a review and evaluation*, NUREC/CR 3649, Office of Nuclear Regulatory Research, Washington, DC.

Stephen, G. A., Brownjohn, J. M. W., and Taylor, C. (1993). "Measurements of static and dynamic displacement from visual monitoring of the Humber bridge." *Journal of Wind Engineering and Industrial Aerodynamics*, 15, 197-208.

Stiemer, S. F., Taylor, P., and Vincent, D. H. C. (1988). "Full-scale dynamic testing of the Annacis bridge." *Proceedings IABSE*, 12(P-122), 1.

Straser, E. G., and Kiremidjian, A. S. (1998). *A modular, wireless damage monitoring system for structures*, 128, John A. Blume Engineering Center, Stanford, CA.

Sumitro, S., Kurokawa, S., Shimano, K., and Wang, M. L. (2005). "Monitoring based maintenance utilizing actual stress sensory technology." *Smart Materials and Structures*, 14, S68-S78.

Talbot, M., and Stoyanoff, S. (2005). "Full-scale modal measurement of the Ile d'Orleans suspension bridge." *Experimental Vibration Analysis for Civil Engineering Structures*, Bordeaux, France.

Tamura, Y., Matsui, M., Pagnini, L. C., Ishibashi, R., and Yoshida, A. (2002). "Measurement of wind-induced response of buildings using RTK-GPS." *Journal of Wind Engineering and Industrial Aerodynamics*, 90(12-15), 1783-1793.

Tanaka, A., Yoshizawa, S., Osawa, Y., and Morisishita, T. (1969). "Period and damping of vibration in actual buildings during earthquakes." *Bulletin of the Earthquake Research Institute*, 47, 1073-1092.

Todd, M. D., Johnson, G. A., Althouse, B. A., and Vohra, S. T. (1998). "Flexural beam-based fiber Bragg grating accelerometers." *Photonics Technology Letters*, 10(11), 1605-1607.

Todd, M. D., Nichols, J. M., Trickey, S. T., Seaver, M., and Nichols, C. J. (2007). "Bragg grating-based fiber optic sensors in structural health monitoring." *Philosophical Transactions of the Royal Society of London, Series A, Mathematical and Physical Sciences*, 365, 317-344.

Trifunac, M. D. (1972). "Comparisons between ambient and forced vibration experiments." *Earthquake Engineering and Structural Dynamics*, 1, 133-150.

Udd, E. (1995). *Fiber optic smart structures*, John Wiley and Sons, New York, NY.

Udwadia, F. E., and Trifunac, M. D. (1974). "Ambient vibration tests on full-scale structures." *5th World Conference on Earthquake Engineering*, Rome.

United States Coastguard and Geodetic Survey. (1936). *Earthquake investigations in California, 1934-1935*, 201, United States Department of Commerce, Washington, DC.

Usher, J., Burch, R. F., and Guralp, C. (1979). "Wide-band feedback seismometers." *Physics of the Earth and Planetary Interiors*, 18(2), 38-50.

Ventura, C. E., Felber, A. J., and Stiemer, S. F. (1996). "Determination of the dynamic characteristics of the Colquitz River bridge by full-scale testing." *Canadian Journal of Civil Engineering*, 23(2), 536-548.

Wemer, S. D., Beck, J. L., and Levine, B. (1987). "Seismic response evaluation of Meloland Road overpass using 1979 imperial valley earthquake records." *Earthquake Engineering and Structural Dynamics*, 15(249), 274.

Wong, K. Y., Man, K. L., and Chan, W. Y. (2001). "Monitoring Hong Kong's bridges: Real-time kinematic spans the gap." *GPS World*, 12(7), 10-18.

Wood, C. J. (1982). *Calibration of a TW 8204 combined wind speed and direction sensor for danjay designs*, 1417/82, Department of Engineering Science, University of Oxford, Oxford, UK.

Worden, K., and Burrows, A. P. (2001). "Optimal sensor placement for fault detection." *Engineering Structures*, 23(8), 685-901.

Wu, Z. S., and Yang, C. Q. (2005). "Self-diagnosis of hybrid CFRP sheets-strengthened structures." *Smart Materials and Structures*, 14, 39-51.

Yanev, B. (2003). "Structural health monitoring as a bridge management tool." *SHM-II: Structural Health Monitoring and Intelligent Infrastructure*, Swets and Zeitlinger,

Yang, C. Q., and Wu, Z. S. (2006). "Self-structural health monitoring function of RC structures with HCFRP." *Journal of Intelligent Material Systems and Structures*, 17(10), 895-906.

Yu, F., and Gupta, N. (2005). "An efficient model for improving performance of vibrating wire instruments." *Measurement*, 37(3), 278-283.

Zasso, A., Vergani, M., Bocciolione, M., and Evans, R. (1993). "Use of a newly designed optometric instrument for long term, long distance monitoring of structures, with an example of its application on the Humber bridge." *2nd International Conference on Bridge Management,* University of Surrey, UK.

Zaurin, R., and Catbas, F. N. (2010). "Integration of computer imaging and sensor data for structural health monitoring of bridges." *Smart Materials and Structures,* 19(1), 1-15.

Zaurin, R. and Catbas, F.N., (2011) "Structural Health Monitoring Using Computer Vision and Influence Lines" *Structural Health Monitoring Journal,* SAGE Publications, Volume 10 Issue 3 May 2011 pp. 309 - 332

Zhang, X., and Brownjohn, J. M. W. (2004). "Effect of relative amplitude on bridge deck flutter." *Journal of Wind Engineering and Industrial Aerodynamics,* 92(6), 493-508.

Chapter 4
Data Processing and Direct Data Interpretation

4.1 Introduction

As many state and federal agencies are developing replacement and rehabilitation strategies for aging infrastructure, it is appropriate and timely to provide support for an integrated condition-assessment framework. This framework should exploit all relevant data in the most effective manner possible. Since low-cost sensor systems are now accessible and reliable, the amount of such data is expected to increase dramatically in the next decade. Structural identification methodologies provide rational and systematic means for data interpretation. Successful data interpretation leads to the following benefits:

- increased efficiency and effectiveness of visual inspection by providing information relating to what to look for and where
- improved decision making for further instrumentation and testing
- better estimations of structural reliability
- better overall structural management for decisions such as replacement planning, retrofit strategies and maintenance budget expenditures
- for civil infrastructure owners and designers, improved insight into what happens to structures during service
- development of an integrated framework for structural condition assessment
- increased generic knowledge of in-service structural behavior that can be distilled into educational materials for students and practitioners
- quantitative contribution to extending concepts of performance-based structural engineering

The first four items of this list are benefits that have the potential to be realized immediately. While the last four benefits are more long term, they stand to have an important impact on the field of structural engineering.

Unfortunately data interpretation challenges often create the final "bottleneck" that restricts the potential of structural identification. In contrast with many other engineering areas, sensors in structural engineering rarely measure causes directly; causes must be inferred from measured effects. Even when causes can be measured in complex structures, it will never be possible to measure directly every possible phenomenon of interest at every location. Thus, without appropriate methods for data interpretation, structural identification cannot provide useful engineering support. Aktan et al. (1997) and Jang et al. (2002) offer comprehensive reviews of early studies of the integration of the analytical and the experimental sides of structural identification. As noted in these and other reviews, there are two main types of data interpretation and they are distinguished by the use or absence of a physics-based behavior model. These two types are complimentary since they are most appropriate in different contexts. Table 4-1 summarizes the strengths and weaknesses of each type.

Table 4-1. Examples of strengths and weaknesses of model-free and model-based data interpretation

Interpretation types	Strengths	Weaknesses
Non-Physics Based (Direct Signal Analysis) Most appropriate when - Many structures need to be monitored - There is time for training the system	• No modeling costs • May not need for damage scenarios • Many options for signal analysis • Incremental training can track damage accumulation • Good for long-term use on structures for early detection of situations requiring model-based interpretation	• Physical interpretation of the signal may be difficult • Weak support for decisions on rehabilitation and repair • Indirect guidance for structural management activities such as inspection and further measurement • Cannot be used to justify replacement avoidance
Physics-based (Structural or Modal Models) Most appropriate when - Design model is not accurate - Structure has strategic importance - Damage is suspected - There are structural management challenges	• Interpretation is easy when links between measurements and potential causes are explicit • The effects of changes in loading and use can be predicted • Guidance for further inspection and measurement • Consequences of future damage can be estimated • Support for planning rehabilitation and repair • May help justify replacement avoidance	• Modeling is expensive and time consuming • Errors in models and in measurements can lead to identification of the wrong model • Large numbers of candidate models are hard to manage • Identification of the right model could require several interpretation - measurement cycles • Complex structures with many elements have combinatorial challenges

Non-parametric models are defined as non-physics-based numerical models that in some cases allow data condensation and reconstruction using a limited number of parameters. This approach does not require the development and use of a behavior model of the structure. Therefore, they are much less onerous to implement. Consequently, they have potential to be used on a large number of structures. For this type of model, the structural identification process is generally a parametric curve-fit of mathematical functions to the measured data. Although the functions reproduce to a certain level of accuracy the measured data, the parameters themselves do not have any direct physical interpretation. The primary goal of this approach is to detect

STRUCTURAL IDENTIFICATION OF CONSTRUCTED SYSTEMS 67

anomalies in behavior. Anomalies are detected as a difference in measurements with respect to measurements recorded during an initial period (Barnett and Lewis 1998). More specifically, this approach involves examining changes over a certain period during the life of a structure. The methodology is data driven in the sense that the evolution of the data is estimated without information of physical processes.

4.2 Examples of Non-Physical Numerical Models

Examples of the data-driven models include autoregressive models (AR) (and variants such as ARMA, ARX and ARMAX models) and the rational polynomial model. In the case of the rational polynomial model, it is interesting to note that although the polynomial coefficients themselves have no particular physical meaning, they can easily be converted to the form of a modal model. Which of these model forms is selected for a particular application depends upon the type of data available, the optimization algorithm to be used, and how the resulting parameters are to be interpreted. With regard to the non-physical numerical model, it can be linear or nonlinear, time-invariant or time varying, and deterministic or probabilistic. This chapter focuses on techniques used for direct data processing and interpretation, without the need for physics-based or parametric models and with specific emphasis on methods ultimately applied to constructed systems. The following approaches are organized by their primary function: **anomaly detection and data processing, data reduction and representation**, and **feature extraction**.

4.2.1 Anomaly Detection

Anomaly detection is an important consideration for any long term monitoring effort. While structural health monitoring has particular interest in detecting anomalies associated with a change to the constructed system (indicative of damage), anomalies in measured signals can also arise for reasons tied to noise, operational and environmental variations of a structure, interference or sensor malfunction. Thus long term monitoring efforts often implement a variety of automated processing measures to remove electrical spikes, sensor drifts and other distortions, which include digital filtering and local spline fitting/interpolation (T. Kijewski-Correa et al. 2006). The sophistication of these methods can be enhanced through the use of statistical significance tests that identify anomalies whenever metrics exceed thresholds established during some training period on the undamaged/new structure. For example, the Instance-Based Method (IBM) (Box and Jenkins 1970; Crawley and O'Donnell 1986; Leontaritis and Billings 1985; Masri et al. 1982) consists of calculating, at each step, the minimum distance of measurements from a neighboring group of sensors (normally 3 or 4) from the cloud of points in the training set. Correlation analyses can also be conducted to monitor sensor outputs for significant variations in measured responses with time that may be indicative of damage. While the total correlation value should be constant or stationary in normal conditions, when damage occurs these values change (Posenato et al. 2008). Correlations are calculated using a moving window selected to guarantee stability of average values, while ensuring rapid damage identification and reducing the effects of noise. Anomalous behavior is quantified by the extent to which these time-evolving correlations fall outside the thresholds defined during the training period.

4.2.2 Data Reduction and Representation

Since it is difficult to characterize data in a high dimensional space, it is often necessary to extract low dimensional features for data analysis. Anomaly detection and data processing may be further augmented by a variety of frequency, time, and time-frequency approaches that enable data to be reduced by a fixed number of parameters. This enables the dominant modes or components to be easily recognized and thus aids in data storage and signal reconstruction, particularly in the reduction of noise. Many of these methods, and the compact representations they offer, further allow the dynamics of the system to be characterized, including potential damage.

4.2.2.1 Frequency Domain and Beyond

One of the most basic transforms applied in the analysis of data is the Fourier Transform, whose subsequent interpretation in auto or cross power spectral densities enables a basic representation of the energy associated with the various modes contributing to measured data. Unfortunately, the fact that data is often characterized by nonstationary or nonlinear features obscured by the harmonic bases of the Fourier Transform prompted a departure from this classical approach. In such cases, it becomes necessary to move to another analysis domain governed by transforms whose bases are compactly supported and produce a time-frequency distribution of energy, discussed in further detail in popular texts by Chui (1992) and Daubechies (1992).

To overcome the limitations of Fourier Transforms, a Short-Time Fourier Transform (STFT) was developed by introducing a short-duration window $w(t)$ centered at time, τ, to the harmonic bases of the original transform. The spectral coefficients could then be determined over this short length of data, which is assumed to be stationary, yielding the STFT. This new breed of transform was one of the first time-frequency distributions, named for their ability to depict energy densities as a dual function of frequency and time through the *spectrogram*. The performance of this transform is greatly dependent upon the choice of window function, which can simply be the traditional boxcar (rectangular) window or more appropriate windows with better performance in both the time and frequency domains (Hanning, Hamming and Gaussian).

Unfortunately, all forms of spectral analyses are limited by the fact that high resolution cannot be obtained in both the time and frequency domains (*Heisenberg Uncertainty Principle*). One hallmark of Fourier and Short-Time Fourier Transforms is the fact that their frequency resolution is fixed throughout, making them ill-suited for analysis of signals that may have both low and high frequency components. This motivated an alternative approach using basis functions with compact support in both frequency and time and then scaled via dilations to optimally adjust their resolutions based on the frequency being analyzed, yielding a multi-resolution analysis. The Wavelet Transform (WT) was engineered with this in mind. The continuous Wavelet Transform (CWT) is a linear transform that generates a time-frequency energy

STRUCTURAL IDENTIFICATION OF CONSTRUCTED SYSTEMS   69

density called a *scalogram* by decomposing a signal $x(t)$ via basis functions that are simply dilations and translations of the parent wavelet $g(t)$ through the convolution with the signal. Dilation by the scale, a, inversely proportional to frequency, represents the periodic or harmonic nature of the signal. Of particular note is the fact that any function satisfying basic admissibility conditions can serve as a parent wavelet. This has served as both an asset (parent wavelets can be chosen that seek a specific characteristic in the signal) and a liability (quality of results is entirely dependent on the parent wavelet selected) of this transform. One of the most common wavelets used in civil engineering applications is the Morlet wavelet, which effectively has a Gaussian-windowed Fourier basis. This transform has been used widely in civil engineering for the analysis of both input excitations and structural response (Gurley and Kareem 1999). Specific to structural identification on constructed facilities, wavelets have been used to observe nonstationary full-scale response features in tall buildings (Bentz and Kijewski-Correa 2009; T. Kijewski-Correa and Pirnia 2007), multipath distortions in full-scale GPS displacement measurements of buildings (T. Kijewski-Correa and Kochly 2007), potential damage of buildings in earthquakes (Hou et al. 2000; T. Kijewski-Correa and Kareem 2004), and the dynamic properties of a tower in Japan (Kijewski and Kareem 2003).

Owing to the sensitivity of wavelets to the parent wavelet chosen, alternate time-frequency transforms have been proposed that are completely empirical in nature. One of the most popular, dubbed the Hilbert Huang Transform (HHT), uses empirical mode decomposition (EMD) to derive narrowbanded intrinsic mode functions (IMFs) that can then be processed as monocomponent signals by the Hilbert Transform (Huang and et al. 1998). These IMFs (c_i) then form the empirical bases of the transform used to reconstruct the signal. The Hilbert Spectrum can then be constructed by Hilbert Transforming each IMF and plotting its squared amplitude as a function of time and frequency, defined by the instantaneous frequency extracted from the phase of the Hilbert Transformed IMF. Similarities and fundamental differences in the representations that result from both Wavelet and Hilbert Huang Transforms have been investigated and debate still surrounds the appropriateness of either transform in certain settings (T. Kijewski-Correa and Kareem 2006; T. Kijewski-Correa and Kareem 2007). Still, HHT has become a popular data-driven transform applied to signal analysis in a variety of fields and even proposed for damage detection in buildings (Pines and Salvino 2006). Specific applications to constructed facilities include parameter identification of the Tsing Ma Bridge (Chen et al. 2004) and the Di Wang Building (Li and Wu 2007).

4.2.2.2 Autoregressive Methods

In light of the aforementioned resolution issues, anomaly detection or feature extraction is often conducted strictly in the time domain using a set of algebraic and temporal relationships among outputs, and in some cases inputs, of systems. Such relationships are useful for predicting values of sensor measurements from measurements of other sensors. Predicted values are then compared with the measured values from those sensors. A temporal redundancy is obtained observing how the differential or difference relationships among different sensor outputs and inputs evolve with the time. A simple relationship for characterizing a system is a polynomial mapping between system inputs (when available) and outputs. One such representation referred to as an autoregressive-moving average with noise, characterizes the system as a weighted polynomial of past outputs (autoregressive - AR) (H. Sohn et al. 2000) and past and present inputs (moving averages - MA). The output is a linear combination of the input history and the past outputs. The input series is a causal moving average (MA) feed-through process, and the series involving weighted past output values is an autoregressive (AR) process. AR (single and multi-variate), ARMA and ARX representations have all been used to represent measured responses of structures, with their performance recently compared in (Su and Kijewski-Correa 2007). In some cases, a linear representation of a process may not be adequate, and the model may have to be extended to include the effects of non-linearities (Loh and Duh 1996). For most practical applications, a polynomial approximation is sufficient. This approach is called polynomial ARX (NARX) (Palumbo and Piroddi 2001; Safak 1991).

In situations where the behavior of the structure varies, it is possible to calculate coefficients incrementally (Brownjohn et al. 2004). This approach is also useful for assessing whether significant information regarding new events can be obtained through observing the autoregressive model (Omenzetter and Brownjohn 2004; Omenzetter et al. 2004). For example, Omenzetter and Brownjohn (2006) investigated how autoregressive methods track time histories of static, hourly sampled strains recorded by an SHM system installed in a major bridge structure and operating continuously for long periods. A seasonal autoregressive integrated moving average (ARIMA) method was established for interpreting recorded strains. Through observing changes in coefficients, unusual events such as sudden foundation settlement, ground movement, excessive traffic loading and failure of post-tensioning cables, can be revealed. In fact, evaluations of the model coefficients and residual errors against baseline values for the structure are also capable of detecting anomalies associated with even minor levels of damage, as first suggested in H. Sohn and Farrar (2001) using a two stage autoregressive approach, and later expanded by a number of authors (T. Kijewski-Correa and Su 2009; Nair et al. 2006) in applications to the ASCE Benchmark Structure and other experimental and simulated datasets.

4.2.2.3 Data Mining

Data mining (Tan et al. 2006; Witten and Frank 2005) is a field of research concerned with finding patterns in data for both understanding and prediction purposes. Data mining algorithms are especially useful when dealing with amounts of data that are so considerable that human processing is infeasible. This is often the situation in structural identification tasks, as visualizing distributions of models in multi-dimensional parameter spaces is difficult for engineers without suitable computing tools. For example, group classification, analysis of outliers, and regressional analyses are all useful for evaluating time evolving data to determine if changes in a structure have occurred (damage) (Gul and Catbas 2009).

Principal Components Analysis (PCA) is an example of a linear data reduction tool that is capable of compressing data and reducing its dimensionality so that essential information is retained and made easier to analyze than the original data set. The main objective is to transform a number of related process variables to a smaller set of uncorrelated variables (Hubert and Verbovney 2003). A key step is finding those principal components that contain most of the information. PCA is based on an orthogonal decomposition of the covariance matrix of the process variables along directions that explain the maximum variation of the data, usually contained in only the first few principal components. As such PCA has considerable value for reconstructing data from the first few principal components to reduce measurement noise. Interestingly, Posenato et al. (2008) found that for a two span continuous beam that is subjected to daily and yearly temperature cycles, moving principal component analysis and moving correlation analysis perform better than wavelet methods, STFT analysis, and the instance-based method.

4.2.3 Feature Selection and Extraction

Feature selection (Dash and Liu 1997) is a method used to reduce the number of features (parameters) in order to facilitate data interpretation. Irrelevant features may have negative effects on a prediction task. Moreover, the computational complexity of a classification algorithm may suffer from excessive dimensionality caused by several features. When a data set has too many irrelevant variables and only a few examples, over-fitting is likely to occur. In addition, data are usually better characterized using fewer variables (Cheng et al. 2007). A comprehensive introduction to feature selection can be found in Guyon and Elisseff (2003). In particular, Feature selection is an effective method for supporting system identification since it identifies parameters that explain predictions of candidate models.

Feature selection techniques (Dash and Liu 1997) can be classified into three main categories (Tan et al. 2006): *embedded approaches* (feature selection is a part of the classification algorithm, i.e., decision tree), *filter approaches* (features are selected before the classification algorithm is used) and *wrapper approaches* (the classification algorithm is used to find the best subset of attributes). Due to its very definition, embedded approaches are limited since they only suit a particular classification algorithm (Molina et al. 2002). Filter methods, however, make the assumption that the feature selection process is independent of the classification step. The work done by Kohavi and Sommerfield (1995) recommends replacement of the filter approach by wrappers, due to the superiority of results, albeit with some computational expense (Weston et al. 2001) and the universality of this approach. Since wrapper techniques treat the classification algorithm as a black box, any search strategy can be used in combination (Kohavi and John 1998). As with any classification algorithm, wrapper feature selection techniques face the over-fitting problem that may happen while training. One way to reduce the overfitting problem is to use a k-fold cross-validation strategy (Bradley and Fayyad 1998; Hsu et al. 2003; Stone 1974).

As noted in Francois (2007), classification techniques such as artificial neural networks (ANN) and support vector machines (SVM) using a Gaussian kernel consider each feature to have equal importance. For this reason, SVM may perform badly when there are many irrelevant features (Weston et al. 2001). The literature contains several applications of SVM to feature selection, with various refinements to enhance performance (Evgeniou et al. 2003; Guyon et al. 2002; Hermes and Buhmann 2000; Liu and Zheng 2006). In Rakotomamonjy (2003), although SVM is used for feature selection, the feature selection and learning process are distinct. Saitta et al. (2010) proposed a feature selection method for structural identification that combines SVM with a global search method (Raphael and Smith 2003) in a wrapper approach. Using a test case of a deck-stiffened arch bridge, it was found that the approach used fewer features to achieve the same accuracy of other approaches that used genetic algorithms.

4.3 Closing Remarks

Over the past ten years, much progress in data interpretation has been made. Advances in enabling technologies, such as computer hardware and diverse arrays of sensors, as discussed in Chapter 3, have provided opportunities to test and evaluate data interpretation methodologies on a scale that was not previously possible. The field has moved from being a part of dynamic system identification, relevant for a small subset of structures, to a key component of an integrated framework for structural condition assessment that is applicable, potentially, to all structures.

Many challenges remain. Further progress (and further challenges) will arise in large part through experience from practical implementations of the proposals described in this report. Specialists need to encourage practitioners to take advantage of new possibilities so that development continues. For example, current AASHTO design practice assumes that bridges are designed on an elemental basis. In addition, AASHTO code specifies that each structural element is to be designed for the loads it will experience during the life of the bridge with probabilistic considerations embedded in load and resistance factors. In reality, loading and the responses of structures may not necessarily conform to the design assumptions. As a result, a more holistic approach is to be executed using experimental data and data interpretation, including explicit representation of important modeling assumptions as parameters within extended sets of models. This is an important aspect of the structural condition assessment framework that is needed for effective and efficient management of civil infrastructure. With this in mind, the next chapter addresses the issues of model selection, model-based identification, and model calibration for structural identification and subsequent decision-making.

4.4 References

Aktan, A. E., Farhey, D. N., Helmicki, A. J., Brown, D. L., Hunt, V. J., Lee, K. L., and Levi, A. (1997). "Structural identification for condition assessment: Experimental arts." *Journal of Structural Engineering,* 123(12), 1674-1684.

Barnett, V., and Lewis, T. (1998). *Outliers in statistical data,* John Wiley & Sons Ltd, West Sussex, UK.

Bentz, A., and Kijewski-Correa, T. (2009). "A wavelet-based framework for system identification of tall buildings under transient wind events." *Proceedings of the 2009 Joint ASCE-ASME-SES Conference on Mechanics and Materials,* Blacksburg, VA.

Box, G. E. P., and Jenkins, G. M. (1970). *Time series analysis, forecasting, and control,* Holden-Day, San Francisco, CA.

Bradley, P. S., and Fayyad, U. M. (1998). "Refining initial points for K-means clustering." *15th International Conference on Machine Learning,* San Francisco, CA.

Brownjohn, J. M. W., Moyo, P., Omenzetter, P., and Chakraboorty, S. (2004). "Interpreting data from bridge performance and health monitoring systems." *International Association for Bridge and Structural Engineering,* Kyoto, Japan.

Chen, J., Xu, Y. L., and Zhang, R. C. (2004). "Modal parameter identification of tsing ma suspension bridge under typhoon victor: EMD-HT method." *J.Wind Eng.Ind.Aerodyn.,* 92(10), 805-827.

Cheng, H., Chen, H., Jiang, G., and Yoshihira, K. (2007). "Nonlinear feature selection by relevance feature vector machine." *Machine learning and data mining in pattern recognition, lecture notes in computer science,* P. Perner, ed., Springer, Berlin/Heidelberg, 144-159.

Chui, C. K. (1992). *Introduction to wavelets,* Academic Press, San Diego, CA.

Crawley, E. F., and O'Donnell, K. J. (1986). "Identification of nonlinear system parameters in joints using the force-state mapping technique." *American Institute of Aeronautics and Astronautics Journal,* 24(1), 155-162.

Dash, M., and Liu, H. (1997). "Feature selection for classification." *Intelligent Data Analysis,* 1(3), 131-156.

Daubechies, I. (1992). *Ten lectures on wavelets,* Society for Industrial and Applied Mechanics, Philadelphia.

Evgeniou, T., Pontil, M., Papageorgiou, C., and Poggio, T. (2003). "Image representations and feature selection for multimedia database search." *IEEE Transactions on Knowledge and Data Engineering,* 15(4), 911-920.

Francois, D. (2007). "High-dimensional data analysis: Optimal metrics and feature selection." PhD Dissertation, Universite Catholique de Louvain, Louvain-la-Neuve, Belgium.

Gul, M., and Catbas, F. N. (2009). "Statistical pattern recognition using time series modeling for structural health monitoring: Theory and experimental verifications." *Mechanical Systems and Signal Processing,* 23(7), 2192-2204.

Gurley, K., and Kareem, A. (1999). "Applications of wavelet transform in earthquake, wind and ocean engineering." *Engineering Structures,* 21(2), 149-167.

Guyon, I., and Elisseff, A. (2003). "An introduction to variable and feature selection." *Journal of Machine Learning Research,* 3, 1157-1182.

Guyon, I., Weston, J., Barnhill, S., and Vapnik, V. (2002). "Gene selection for cancer classification using support vector machines." *Journal of Machine Learning Research,* 46(1-3), 389-422.

Hermes, L., and Buhmann, J. M. (2000). "Feature selection for support vector machines." *15th International Conference on Pattern Recognition,*

Hou, Z. K., Noori, M., and Amand, R. S. (2000). "Wavelet-based approach for structural damage detection." *ASCE Journal of Engineering Mechanics,* 126(7), 677-683.

Hsu, C. W., Chang, C. C., and Lin, C. J. (2003). *A practical guide to support vector classification,* National Taiwan University, Taipei City, Taiwan.

Huang, N. E., and et al. (1998). "The empirical mode decomposition and the Hilbert spectrum for non-linear and non-stationary time series analysis." *Philosophical Transactions of the Royal Society of London, Series A, Mathematical and Physical Sciences,* 454, 903-995.

Hubert, M., and Verbovney, S. (2003). "A robust PCR method for high-dimensional regressors." *Journal of Chemometrics,* 17(438), 452.

Jang, J. H., Yeo, I., Shin, S., and Chang, S. P. (2002). "Experimental investigation of system-identification-based damage assessment on structures." *ASCE Journal of Structural Engineering,* 128(5), 673-682.

Kijewski, T., and Kareem, A. (2003). "Wavelet transforms for system identification in civil engineering." *Computer-Aided Civil and Infrastructure Engineering,* 18, 339-355.

Kijewski-Correa, T., and Kareem, A. (2004). "Time-frequency perspectives in the analysis and interpretation of ground motions and structural response." *Proceedings of 13th World Conference on Earthquake Engineering,* Vancouver, BC.

Kijewski-Correa, T., Kilpatrick, J., Kareem, A., Kwon, D. K., Bashor, B., Kochly, M., Young, B. S., Abdelrazaq, A., Galsworth, J. K., Isyumov, N., Morrish, D., Sinn, R. C., and Baker, W. F. (2006). "Validating wind-induced response of tall buildings: Synopsis of the Chicago full-scale monitoring program." *ASCE Journal of Structural Engineering,* 132(10), 1509-1523.

Kijewski-Correa, T., and Kochly, M. (2007). "Monitoring the wind-induced response of tall buildings: GPS performance and the issue of multipath effects." *Journal of Wind Engineering and Industrial Aerodynamics*, 95(9-11), 1176-1198.

Kijewski-Correa, T., and Pirnia, J. D. (2007). "Dynamic behavior of tall buildings under wind: Insights from full-scale monitoring." *Structural Design of Tall and Special Buildings*, 16, 471-486.

Kijewski-Correa, T., and Su, S. (2009). "BRAIN: A bivariate data-driven approach to damage detection in multi-scale wireless sensor networks." *Smart Structures and Systems*, 5(4), 415-426.

Kijewski-Correa, T., and Kareem, A. (2006). "Efficacy of Hilbert and wavelet transforms for time-frequency analysis." *J.Eng.Mech.*, 132(10), 1037-1049.

Kijewski-Correa, T., and Kareem, A. (2007). "Nonlinear signal analysis: Time-frequency perspectives." *J.Engrg.Mech.*, 133(2), 238-245.

Kohavi, R., and John, G. (1998). "The wrapper approach." *Feature extraction construction and selection: A data mining perspective*, H. Liu and H. Motoda, eds., Kluwer Academic Publishers, Norwell, MA, 33-50.

Kohavi, R., and Sommerfield, D. (1995). "Feature subset selection using the wrapper model: Overfitting and dynamic search space topology." *1st International Conference on Knowledge Discovery and Data Mining*,

Leontaritis, I. J., and Billings, S. A. (1985). "Input-output parametric models for nonlinear systems, part I: Determininistic nonlinear systems." *International Journal of Control*, 41(3.2, 6.3), 303-328.

Li, Q. S., and Wu, J. R. (2007). "Time-frequency analysis of typhoon effects on a 79 storey building." *Journal of Wind Engineering and Industrial Aerodynamics*, 95, 1648-1666.

Liu, Y., and Zheng, Y. F. (2006). "FS_SFS: A novel feature selection method for support vector machines." *Pattern Recognition*, 39(7), 1333-1345.

Loh, C. H., and Duh, J. Y. (1996). "Analysis of nonlinear system using NARMA models." *JSCE Journal*, 13, 11-21.

Masri, S. F., Bekey, G. A., Sassi, H., and Caughey, T. K. (1982). "Non-parametric identification of a class of non-linear multidegree dynamic systems." *Earthquake Engineering and Structural Dynamics*, 10, 1-30.

Molina, L. C., Belanche, L., and Nebot, A. (2002). "Feature selection algorithms: A survey and experimental evaluation." *IEEE International Conference on Data Mining*, Maebashi City, Japan.

Nair, K. K., Kiremidjian, A. S., and Law, K. H. (2006). "Time series based damage detection and localization algorithm with application to the ASCE benchmark structure." *Journal of Sound and Vibration*, 291(1), 349-368.

Omenzetter, P., and Brownjohn, J. M. W. (2004). "Application of time series and kalian filtering for structural health monitoring of a bridge." *2nd European Workshop on Structural Health Monitoring*, Munich, Germany.

Omenzetter, P., and Brownjohn, J. M. W. (2006). "Application of time series analysis for bridge health monitoring." *Smart Materials and Structures*, 15, 129-138.

Omenzetter, P., Brownjohn, J. M. W., and Moyo, P. (2004). "Identification of unusual events in multi-channel bridge monitoring data." *Mechanical Systems and Signal Processing*, 18, 409-430.

Palumbo, P., and Piroddi, L. (2001). "Seismic behavior of buttress dams: Nonlinear modeling of a damaged buttress based on ARX/NARX models." *Journal of Sound and Vibration*, 239, 405-422.

Pines, D., and Salvino, L. (2006). "Structural health monitoring using empirical mode decomposition and the Hilbert phase." *J.Sound Vibrat.*, 294(1-2), 97-124.

Posenato, D., Lanata, F., Inaudi, D., and Smith, I. F. C. (2008). "Model-free data interpretation for continuous monitoring of complex structures." *Advanced Engineering Informatics*, 22(1), 135-144.

Rakotomamonjy, A. (2003). "Variable selection using SVM-based criteria." *Journal of Machine Learning Research*, 3, 1357-1370.

Raphael, B., and Smith, I. F. C. (2003). "A direct stochastic algorithm for global search." *Journal of Applied Mathematics and Computation*, 146(2-3), 729-758.

Safak, E. (1991). "Identification of linear structures using discrete-time filters." *Journal of Structural Engineering*, 117(10), 3064-3085.

Saitta, S., Kripakaran, P., Raphael, B., and Smith, I. F. C. (2010). "Feature selection using stochastic search: An application to system identification." *Journal of Computing in Civil Engineering*, 24(1), 3-10.

Sohn, H., Czarneski, A., and Farrar, C. R. (2000). "Structural health monitoring using statistical process control." *Journal of Structural Engineering*, 126(11), 1356-1363.

Sohn, H., and Farrar, C. R. (2001). "Damage diagnosis using time series analysis of vibration signals." *Smart Materials Structures,* 10(3), 446.

Stone, M. (1974). "Cross-validatory choice and assessment of statistical predictions." *Journal of the Royal Statistical Society,* 36(2), 111-147.

Su, S., and Kijewski-Correa, T. (2007). "Performance verification of bivariate regressive adaptive index for structural health monitoring." *SPIE's 12th Annual International Symposium on NDE for Health Monitoring and Diagnostics,* SPIE, San Diego, CA.

Tan, P. N., Steinbach, M., and Kumar, V. (2006). *Introduction to data mining,* Addison-Wesley, Boston, MA.

Weston, J., Mukherjee, S., Chapelle, O., Pontil, M., Poggio, T., and Vapnik, V. (2001). "Feature selection for SVMs." *Advances in neural information processing systems 13,* T. K. Leen, T. G. Dietterich and V. Tresp, eds., MIT Press, Cambridge, MA, 1.

Witten, I., and Frank, E. (2005). *Data mining: Practical machine learning tools and techniques with java implementations,* Morgan Kaufman Publishers, San Francisco, CA.

Chapter 5
Structural Identification for Selection, Application, and Calibration of Physics-Based Models

5.1 Model Selection

Methods for structural identification differ in terms of model form and space. Model form refers to the mathematical expression that will be used to predict the structural behavior. Model space refers to the type of coordinate system (i.e. related to the degrees of freedom) that defines the model. Model form and space are closely interrelated and therefore they are usually defined together to describe the structural identification process.

5.1.1 Structural Model

The structural model, discussed previously in Chapter 2 and specifically the finite element model, provides structural connectivity and property information in the form of elemental force-displacement relationships and material constitutive properties. The structural model typically requires assumptions of the linearity and time-invariance of model parameters; however, this is not a general limitation. For example, a finite element model can include nonlinear constitutive relations for a material. Additionally, structural models are typically defined deterministically, but may also be defined in a probabilistic sense. For example, in stochastic finite element analysis material properties can be defined in terms of a probability distribution, i.e., the material property is known statistically within a certain distribution function.

5.1.1.1 Use of a Single Model

The vast majority of research studies and most practical applications have involved the use of a single model. It is common practice to assume that the service behavior of a structure can be modeled using the same model that was used in design if one was created. Since this assumption requires no verification of characteristics such as support conditions, geometry, and damage on the as-built structure, it is clearly the most attractive in terms of time and money. The predictions of this model are then compared with measurements. When agreement between measurement data and model predictions is not satisfactory, measurements are used to "calibrate" the model. This involves selecting a small number of model parameters that may have values that are different from those used in design. It is more desirable to select the parameters that have uncertainty. For example, it is much easier to justify the updating of a boundary condition parameter than the Young's Modulus of steel. Once these parameters are selected, various procedures are used to find their values for which the measurements best match the model predictions, as discussed later in this chapter. Calibration and validation are two separate tasks. Calibration of a model that has not been validated is likely to lead to results that do not represent real structure behavior.

5.1.1.2 Use of Multiple Models

Especially after decades of service life, many behavior models may potentially be accurate descriptions of real behavior, and within each model, many (sometimes thousands) of combinations of parameter values may provide reasonable explanations of measured behavior. In such cases, the structural identification challenge is even further complicated, especially since most methods do not explicitly consider the combined effects of modeling and measurement error. Most work to date involves manual selection of a few likely models using engineering experience, each differing in their treatment of certain properties such as member stiffness, support (boundary) conditions, and use of in-span hinges. In other cases, the use of a data mining method such as clustering can give engineers an idea of the topology of the candidate model space (Saitta et al. 2008). The methodology helps engineers by providing cluster centers as possible models that explain the structural behavior. Meanwhile, (Robert-Nicoud, Raphael, and Smith 2005) proposed a multiple-model identification methodology based on compositional modeling and a stochastic global search to generate a set of candidate models that are within a root-mean-square (RMS) error of the difference between measured values and model predictions. Note that a model is a candidate model only if it satisfies the RMS error condition at each measurement location, i.e., the difference is within the specified threshold for every single measurement location (Robert-Nicoud, Raphael, Burdet et al. 2005).

5.1.2 Modal Model

Another model form commonly used in structural identification is the modal model. This model consists of modal frequencies, modal vectors (also called mode shapes) and modal damping ratios. Together these components are referred to as modal parameters. The modal model is different from the structural model in that it does not contain specific information about the structural connectivity or the geometric distribution of mass, structural damping, and stiffness. However, a modal model is useful for several reasons. First, the modal parameters that form the parametric description of the modal model also describe the resonant spatial and temporal behavior of the structure. Therefore, they are intuitive to the practicing engineer as a convenient form for expression of the structural behavior. Second, the modal parameters are directly analogous to the eigensolution of the structural mass and stiffness matrix. This parallel between the experimentally derived and analytically derived components makes the modal model well suited for the process of model correlation. Also, the modal model is convenient because the structural frequency response function can be written in canonical form in terms of the modal parameters. For response simulations in the linear range, a modal model of reliable accuracy is all that is required for performance calculations for a range of loading scenarios. This is now routinely used for floors, grandstands and footbridges for human dynamic load cases (Pavic et al. 2010). Also the structural frequency response function can be measured using standard experimental techniques, such as those detailed in Chapter 3, so the modal parameters can be identified from the measured data using a number of different algorithms discussed later in this chapter.

The modal model is defined in a coordinate space known as modal coordinates. These coordinates form a generalized basis for describing the vibratory motion of the structure with a relatively small number of parameters. In modal coordinates, the equations of motion for an undamped system can be written as

$$[m_r]\{\ddot{q}(t)\} + [k_r]\{q(t)\} = \{f_r(t)\} \tag{1}$$

where $[m_r]$ and $[k_r]$ are the modal mass and stiffness matrices, respectively, $\{q(t)\}$ are the modal displacements, and $\{f_r(t)\}$ are the modal forces. The response of this equation to excitation with a sinusoid at frequency ω can be written as

$$\alpha_{jk}(\omega) = \sum_{r=1}^{2N} \frac{{}_r A_{jk}}{\omega_r \zeta_r + i(\omega - \omega_r')} \tag{2}$$

This equation is complex-valued and expresses a magnitude and phase of the response at each excitation frequency relative to the magnitude and phase of the excitation sinusoid. The modal model is inherently linear and time invariant, so it is assumed that the structure to be identified conforms to those assumptions as well. Typically, the modal parameters are defined only deterministically; however, there is an increasing attention in the literature to the expression of modal parameters in stochastic terms. In fact, because of variability in environmental conditions, test repeatability, and other random and systematic factors, the measurements made for structural identification are always associated with some level of uncertainty. The definition of such stochastic models allows the engineer to define the level of certainty with which the modal parameter values are known. Stochastic representations of models usually assume that parameters vary independently of one another. Mutual correlation and exclusion are rarely included. As a result, stochastic representations are usually rough estimations of the probability space. It is debatable whether such increases in sophistication are practically useful.

5.2 Model Application for Structural Identification

Once the model form and space has been determined, an appropriate technique must be selected to identify the parameters of that model. This selection depends on a variety of factors, summarized in Table 5-1, including the details of the instrumented degrees of freedom (sensor density and type), the availability of measured input excitation, and the nature of the excitation to the system. For the last three decades, numerous structural identification methods have been developed for dynamically loaded structures under these wide ranging conditions commonly observed in the fields of aerospace, mechanical and structural engineering; specific focus on those methods that have been applied to constructed facilities is provided next.

Most methods of parameter identification are essentially regression techniques that estimate the parameters in models to simulate physical systems based on the outputs of (and in some cases input to) the systems using various optimization algorithms. Some of the classical approaches of parameter estimation include weighted least-squares estimation, best linear unbiased estimation (BLUE), maximum likelihood for deterministic parameters, mean squared, maximum a posteriori, weighted least-squares and BLUE for random parameters (Mendel 1995). More recently, the extended or unscented Kalman filter (Ghanem and Shinozuka 1995; Hoshiya and Saito 1984; Wu and Smyth 2007; Yun and Shinozuka 1980; Zhang et al. 2002), H∞ filter (Sato and Qi 1998; Sato and Qi 1999), sequential Monte Carlo methods (Ching et al. 2006; Li et al. 2004; Yoshida 2001) and regression techniques based on support vector machines (Mita and Hagiwara 2003; Oh and Beck 2006; Worden and Lane 2001) have been used in parameter identification for constructed systems. The following sections will now showcase methods commonly used in structural identification, based on the type of underlying model they assume.

Table 5-1. Summary of aspects of model classification for vibration-based parameter identification

Aspect	Possibilities
Linearity	Linear, nonlinear
Modeling scope	System-level, component-level
	Global vs. local
Mass distribution	Discrete, continuous
	lumped-mass, distributed-mass;
	building-like, bridge-like
Domain of modeling	Time-domain vs. frequency-domain, modal vs. structural
Degree of sequential correlation of data	Stationary, non-stationary
Number of system inputs and outputs	Single-input single-output (SISO) vs. multi-inputs multi-outputs (MIMO) and SIMO
Availability of system inputs	Input-output systems, output-only
	forced vibration, ambient vibration
Uncertainty	Epistemic, aleatory

5.2.1 Identification Using Structural Models

Direct identification of time-varying damping and stiffness matrices is possible, provided the response is measured at all or most of the degrees-of-freedom (DOFs). For example, subspace state-space system identification has been conducted on the 15-story UCLA Building, modeled as a 45-DOF system, utilizing 72 uniaxial accelerometers (Skolnik et al. 2006), with similar applications discussed later in Chapter 7. While this method generally requires both the measured input and output, it can be used for output-only (ambient) identification by expanding the state space model to generate extra "numerical modes" accounting for the unknown input and noise. A 90th order model was used for the UCLA Building for analysis of ambient vibration data, necessitating additional stability analyses. While the response at uninstrumented degrees-of-freedom can be generated by interpolation or state-space observers, as in the case of the Fire Command and Control Building in California (Nagarajaiah and Li 2004), the limited number of sensors in most large-scale applications often prohibits a unique determination of stiffness and damping matrices. Thus, identification using modal models tends to be more popular in practice, as they permit all "visible" modes to be determined uniquely at each measurement point, where the term visible implies that the higher modes are sufficiently excited and not obscured by noise.

Structural identification approaches have also been extended to damage detection, with attempts to apply them to constructed facilities. For example, (Turek and Ventura 2007) applied the Damage Locating Vector (DLV) technique, a flexibility-based method (Bernal and Gunes 2004), which was adapted to the ambient vibration analysis using approximate flexibilities by the Stochastic Damage Locating Vector (SDLV) (Bernal 2006) and the Proportional Flexibility Matrix (PFM) (Duan et al. 2005) techniques. These approaches were employed on the Melville Building and the Heritage Court Tower, respectively 44- and 15-story reinforced concrete buildings in Vancouver. This study highlighted a major practical limitation in flexibility-based identification: a degree-of-freedom mismatch or the difficulty in achieving a sufficient sensor density to resolve even a simplified frame representation of the inherently more complex actual structure. Even more challenging is the fact that in output-only identification only approximate mode shapes can be obtained, leading to a failure of the DLV in the field.

In addition to stiffness and flexibility-based structural identification, derivatives of the stiffness matrix can be used for damage identification (Catbas et al. 2008), as well as more explicit structural identification using NDT data from the structure can also be conducted (Sanayei et al. 2006). This form of parameter estimation procedure can enable adjustment of the model's mass and stiffness parameters such as axial rigidity (EA), flexural rigidity (EI) torsional rigidity (GJ), support stiffness, lumped mass, and element distributed mass per unit length. Techniques for systematic updating of baseline models are discussed later in Section 5.3.

5.2.2 Identification Using Modal Models

As summarized by number of studies (Allemang et al. 1994; Allemang 1995; Farrar et al. 2003; Maia and Silva 1997), methods for modal parameter identification can be categorized according to the order of the data that they use (low or high order) and according to the domain that the data is analyzed in (time or frequency domain), the number of inputs and outputs (Catbas et al. 2004), and even the input excitation (ambient vs. seismic) they are acquired under (Kijewski-Correa and Cycon 2007; Kijewski-Correa et al. 2008). This section summarizes some of the most commonly used modal parameter identification techniques.

5.2.2.1 Complex Exponential Method

The fundamental method among the high-order, time-domain, modal parameter estimation techniques is the Complex Exponential Algorithm (CEA) (Brown et al. 1979; Spitznogle and Quazi 1970; Spitznogle et al. 1971). The fundamental limitation of the basic CEA method is that it can only be applied to one response artifact (impulse response function derived from the frequency response function) at a time. Thus, a complete set of modal parameters is computed for each excitation-response pair, even for global parameters such as modal frequency (which should only have

STRUCTURAL IDENTIFICATION OF CONSTRUCTED SYSTEMS   83

one value over all the DOF for each mode). The analyst may, by inspection, correlate the common modes across all the identification sets, but this is at best a tedious process. To overcome the limitation of the CEA procedure to single input, single output (SISO) problems, the Least-Square Complex exponential algorithm (LSCEA) was developed. LSCEA allows CEA to be applied over multiple response DOF simultaneously to produce an appropriately sized set of global modal parameters to yield a single input multiple output (SIMO) approach (Brown et al. 1979). While the modal frequencies and modal damping ratios are indeed global properties, they are not observed identically from all response DOF as a result of measurement variability, system property variability, etc. Thus, a least-square solution is employed to find the solution that minimizes these errors. The fundamental limitation of LSCEA is that the solution is limited to the SIMO case and will need to be applied repeatedly for MIMO data. The repeated application of the technique leads to the same problems observed in the basic CEA.

A MIMO technique based on the principles of CEA and LSCEA (Vold and Rocklin 1982; Vold et al. 1982) is known as the Polyreference Complex Exponential or Polyreference Time Domain technique (Deblauwe et al. 1985; Maia and Silva 1997), which also has a frequency domain implementation called Polyreference Frequency Domain (PFD) (Zhang et al. 1985). The capability to process MIMO data removed several major obstacles in modal parameter identification up to that point. For one, allowing for multiple excitation locations minimizes the possibility that a particular vibration mode will be unobservable. Also, multiple excitations in the data set enable the resolution of spectrally repeated or closely spaced modes.

Companion methods operating on impulse responses include the Ibrahim Time Domain (ITD) method and the Eigensystem Realization Algorithm (ERA) (Juang and Pappa 1985; Juang 1994), which also has a frequency domain implementation as ERA-FD, or stochastic subspace identification technique (Van Overchee and de Moore 1996), two quite similar first-order time domain modal parameter identification techniques. The primary differences are that ITD is SIMO while ERA is MIMO, with each employing different procedures to solve the overdetermined set of equations produced during the formulation. Autocorrelation functions have been fed into ERA to extract the dynamic properties of a building at Saitama University, Japan (Areemit et al. 2003), applied to the responses of the Pioneer Bridge in Singapore (Brownjohn et al. 2003), and used in conjunction with the natural excitation technique (NExT) (Caicedo et al. 2004; James et al. 1993) for the ASCE Benchmark Structure and the Vincent Thomas and Hakucho Bridges (presented as case studies later in Sections 8.4 and 8.5). (Areemit et al. 2003) underscored the importance of sufficient data to accurately estimate the correlation function, as well as to form a Hankel matrix of sufficient dimension. As commonly observed in other ambient vibration investigations, frequencies were repeatably estimated, though damping and mode shapes showed poor stabilization and were thus deemed unreliable. Similar experiences with direct determination of autocorrelation functions on tall buildings (Davenport and Hill-Carroll 1986) have helped propel the Random Decrement Technique (RDT) as an alternative for autocorrelation estimation.

5.2.2.2 Random Decrement Technique

The lack of measured input in a number of applications, particularly those involving wind-induced excitations, precludes the use of many structural identification approaches. One of the most popular output-only structural identification approach is the Random Decrement Technique, generated by capturing a sample of prescribed length from a stationary time history upon the satisfaction of a threshold condition specified in terms of response amplitude and slope (Cole 1973). The segments meeting these conditions are averaged to remove the random component of the response, assumed to be zero mean, leaving Random Decrement Signature (RDS). The RDS was shown to be proportional to the autocorrelation function (Vandiver et al. 1982), but without the same strict requirements for lengthy stationary data (Jeary 1992) and with the ability to track amplitude dependent dynamic properties (Tamura and Suganuma 1996). If the input process is Gaussian, zero mean white noise, the RDS of a linear system will be proportional to the free vibration response (Gurley and Kareem 1999; Spanos and Zeldin 1998), allowing damping and frequency to be extracted by a number of approaches depending on the response being characterized by a single mode (logarithmic decrement or analytic signal theory) or multiple modes (Ibrahim Time Domain Method).

RDT, with varying refinements, has been widely used in the structural identification of frequency and damping from tall buildings excited by wind. For example, it is one of the preferred methods to estimate frequency and damping from full-scale responses cataloged in databases by Japanese (Satake et al. 2003) and Korean (Yoon and Ju 2004) researchers. In particular, RDT has successfully isolated constant and, with variable trigger amplitudes, amplitude-dependent dynamic properties in a number of tall buildings in China (a 30 story tall building in Hong Kong, DiWang Tower, Bank of China, Central Plaza, Jin Mao Tower, Guangdong International Building), a tall building in Boston, and a tall building in Korea (Fang et al. 1999; Kijewski-Correa et al. 2006; Kijewski-Correa and Pirnia 2007; Q. S. Li et al. 2003; Li et al. 2004; Li et al. 2004; Q. S. Li et al. 2005; Q. S. Li et al. 2006; Pirnia et al. 2007). This method is also used in case studies at the conclusion of this report involving three tall buildings in Chicago (Section 7.1) and a pair of bridges (Section 8.5).

5.2.2.3 Power Spectral Approaches

Analysis of the power spectral density is an important form of structural identification. In particular, the analysis of ambient or wind induced responses, where as little as two sensors are employed, commonly invoke the assumptions of a stationary, ergodic response to enable frequency and damping to be estimated from the power spectral peaks. The identification can be conducted by point estimators, similar to the half power bandwidth (HPBW), applied extensively in the European and Japanese full-scale databases (Lagomarsino and Pagnini 1995; Satake et al. 2003) and the analysis of two tall buildings under typhoons in Hong Kong (Campbell et al. 2005). Still, there is significant reluctance to use frequency domain methods due to the many signal processing issues associated with fast Fourier transforms and the inclusion of windows that inflate damping estimates. In particular, given the increased variance that often results when limited amounts of data are available, least squares fits (Lagomarsino and Pagnini 1995) such as maximum likelihood estimators have been employed and were shown to produce lower bias and variance in their damping estimates than traditional methods when applied to a collection of over 60 buildings in South Korea (Erwin et al. 2007). Alternatively, techniques such as Frequency Domain Decomposition (FDD) (Brincker et al. 2001) employing curve fitting in the vicinity of the singular values has grown increasingly popular for frequency-domain identification, even in the presence of high variance with limited amounts of data. This method was employed for the long term monitoring of the first nine modes of the Tokyo International Airport and New Tokyo International Airport Control Towers using arrays of 27 accelerometers (Yoshida and Tamura 2005). This method proved especially advantageous in the evaluation of closely spaced modes in full-scale (Yoshida et al. 2004). FDD has also been used, however, in this application, the singular values meeting minimum modal assurance criteria were transformed back to the time domain through inverse Fourier transform to permit logarithmic estimates of damping (Rainieri et al. 2007).

5.2.2.4 Applications in Base-Excited Structures

Modal identification applied to recorded seismic response is a rather mature area of system identification research, aided by the availability of the measured input. Most of the studies have estimated the modal properties by using a non-linear optimization to perform time-domain least-squares matching of the measured and model responses based on a model consisting of the superposition of the lower modes (Ljung 1998). Early theory and applications are in publications by Beck (Beck 1978; Beck and Jennings 1980) and, for multi-input multi-output (MIMO) systems, in (Werner et al. 1987); (see also, the Theory section at the website for the Caltech Online Monitoring and Evaluation Testbeds). Least-squares time-domain methods have also been applied to bridges including the Vincent Thomas Bridge, addressed in detail as a case study in Section 8.4 (Masri et al. 1987; Smyth et al. 2003). The least-squares output-error matching can also be performed in the frequency domain using the complex Fourier transforms of the data and model outputs (McVerry 1980); the time-domain and frequency-domain versions are essentially equivalent because of Parseval's identity. Usually the modal models are based on uncoupled classical modes, but least-squares matching based on non-classical modes of vibration can also be used (Tan and Cheng 1993).

For example, the identification problem can be treated as a discrete-time filter design (Safak 1989a). This approach has been applied to an anonymous 22 story building in Chile, Imperial Company and Services Building (California), the Transamerica Building, Pacific Park Plaza, and Bank of California in Los Angeles (Safak 1989b; Safak 1991; Safak and Celebi 1991; Safak and Celebi 1992; Safak 1993) and is adopted in the case studies presented later in Sections 7.3 and 8.3. Due to its wide availability in commercial software packages, regressive time-series modeling of the recorded input and output accelerations using least squares minimization and extraction of dynamic properties from the poles of the resulting transfer function has become popular (Celebi 1993). Various regressive models have been adopted, including AutoRegressive model with eXogenous input (ARX), used for the Atwood Building in Alaska and several buildings in California: Alhambra

86   STRUCTURAL IDENTIFICATION OF CONSTRUCTED SYSTEMS

Building in Los Angeles, an anonymous 47 story building in San Francisco, Pacific Park Plaza, Transamerica Building, Santa Clara County Office Building, San Bruno Office Building and the CSUH Administration Building (Celebi 1993; Celebi 1996; Celebi 2006; Rodgers and Celebi 2006) and the autoregressive moving average model with exogenous input (ARMAX), applied to two steel buildings in Japan, as part of the damping database discussed later (Saito and Yokota 1996).

An output-error minimization of the relative acceleration and the derived relative velocity and displacement, using a modal sweeping scheme for efficient and reliable convergence has been used in applications to the Union Bank Building in Los Angeles (Beck and Jennings 1980). This modal minimization method has been applied to a number of other buildings: the Town Park Towers, the Great Western Savings and Loan Building in San Jose, the Santa Clara Co. Office Building, the Law & Justice Center in Rancho Cucamonga, the San Fernando Holiday Inn, the Santa Monica Sheraton-Universal Hotel and the Hollywood Storage Building (Lin and Papageorgiou 1989; Papageorgiou and Lin 1989a; Papageorgiou and Lin 1989b; Papageorgiou and Lin 1991a; Papageorgiou and Lin 1991b). This type of least squares minimization of the output (acceleration) errors has also been applied to anonymous 13 and 47 story buildings in California (Li and Mau 1991; Loh and Tou 1995). Using the same response quantities for their objective function, though with some normalization, the optimization was approached stochastically via genetic algorithms to accommodate larger numbers of variables and objective function complexity (Alimoradi et al. 2006). This technique was applied to extract time-variant dynamic properties from 68 full-scale response cases from a variety of instrumented buildings (e.g., Millikan Library on the campus of CalTech, Van Nuys Hotel, Imperial County Services Building, and Sylmar County Hospital all located in California) (Alimoradi and Naeim 2006).

These and other past studies of data recorded in low to mid-rise buildings during earthquakes (Beck 1978; Beck and Jennings 1980; Durrani et al. 1994; Hashimoto et al. 1993; Li and Mau 1991; Mau and Aruna 1994; McVerry 1980; McVerry and Beck 1983; Nisar et al. 1992; Papageorgiou and Lin 1989a; Papageorgiou and Lin 1989b) and have revealed that a small number of modes (5 to 10) can capture the translational and torsional behavior of tall buildings during non-damaging earthquakes. For the most part, these efforts have focused on equivalent linear modal parameters from "time-invariant" models of a structure, where the entire duration of the recorded earthquake motions is used. In addition, the time variation of these parameters has been studied by using smaller time windows of data (two fundamental periods or more in length). This procedure can provide insight into the extent and nature of the nonlinearities in the structural behavior. The estimated prediction-error standard deviations, which are given by the output RMS error for the optimal identified modal model, range from about 15% to 70% of the RMS of the measured accelerations, with the low end corresponding to small non-damaging seismic motions and the high end corresponding to damaging response where "time-invariant" linear dynamic models do not capture well the enormous changes in stiffness that occur.

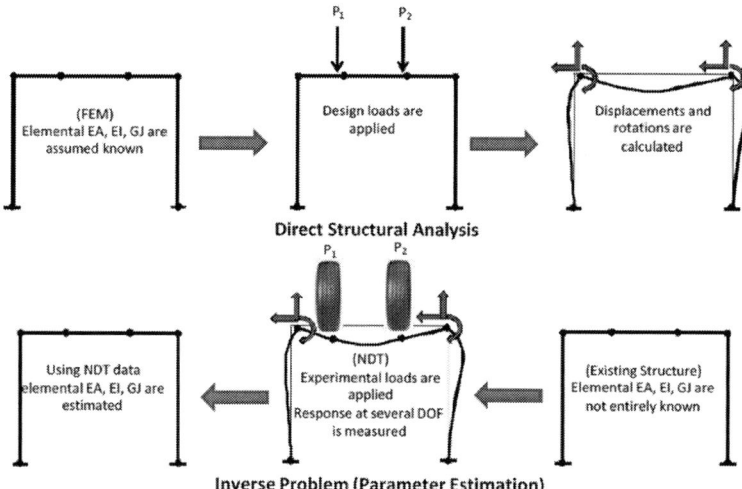

Figure 5-1. Structural identification and inverse analysis (Adapted from Santini Bell et al. 2007).

5.3 Finite Element Model Calibration

Once structural parameters have been identified from experimental data, they can be used to calibrate numerical models so that their response predictions correlate well with the measured response of the physical system, as shown in Figure 5-1. Differences between in-situ and predicted structural parameters and responses may arise from simplifications employed in the modeling process, e.g., in the representations of the boundary and support conditions, connectivity between various structural elements, unknown material properties and constitutive relationships (particularly those associated with soil and concrete), and energy dissipation (damping) mechanisms as well as measurement errors. However, differences may also arise due to changes in the condition of the structure due to factors such as deterioration, support settlement, loss of prestressing forces and changes in end conditions and so far the effects of modeling errors and structural degradation cannot be distinguished by this technology.

Unfortunately, such correlations are not currently part of general civil engineering practice, and the majority of the model correlation studies that have been reported in the civil engineering literature have been part of research efforts or studies of structures with unique features or demands that warrant special investigations, e.g., performance during earthquakes or identification of bridge boundary conditions. The calibration process involves selecting a small number of model parameters that have uncertainty so their values cannot be known a priori. Once these parameters are selected, various procedures are used to find their values for which the measurements best match the model predictions. This process then naturally enables updating the analytical models such that they more accurately predict the observed response of the *in situ* structural system (Aktan et al. 1997; Jang et al. 2002). However, despite the advances in finite element modeling, model calibrations of full-scale structures can easily be in error by as much as 50%, indicating that validation of a particular behavior model is a non-trivial exercise (Catbas et al. 2007). This process can be particularly challenging due to the degree of freedom mismatch, as the number of response measurement locations is significantly less than the number of degrees of freedom in the finite element model. This mismatch often makes it difficult to precisely identify the portions of the model that cause the discrepancies between measured and predicted response. Recent advancements in computational capabilities, have enabled more advanced techniques to overcome these challenges (Guo 2002) presents a method to improve a structural model through use of data to identify poorly modeled regions of a structure. No matter the approach used, the result of a successful calibration effort is a model suitable to provide owners and managers of that infrastructure with the information necessary for decision making related to rehabilitation and maintenance. For example, current load rating of an in-service bridge and prediction of the remaining service life given current loading conditions, generated by inverse engineering that is driven by abductive logic (Raphael and Smith 2003). A calibrated model enables a more pro-active maintenance that can be substantially more economical than delayed responses to deterioration. In fact, the range of utility of structural identification is much wider than the damage-identification perspective that is prevalent in other fields (Worden et al. 2007), making it a valuable asset to the civil engineering community. The next section discusses how errors in numerical models are characterized, objective functions are created, and models are updated.

5.3.1 Error Characterization

Historically, the structural engineering community has taken a component level approach to experimental verification of analysis and design procedures. Individual components such as beams, columns, base isolators, or connections are tested and the results of these tests are used to generate design requirements for the specific component. Meanwhile, system-level structural demands are purposely overestimated and element level structural capacities are purposely underestimated to add conservatism into the design process, as reflected in national structural design codes by AISC and ACI. This conservatism is used to compensate for uncertainties in behavioral prediction and the system modeling.

Modeling error (e_{mod}), defined as the difference between the predicted response of a given model and that of an ideal model, has been studied by many researchers (Arya et al. 1998; Banan and Hjelmstad 1994; Frangopol and Liu 2007; Kong and Frangopol 2003; Sanayei et al. 1997). This modeling error can be characterized in a number of ways. For example, (Liu and Frangopol 2004) consider two types of uncertainties: aleatory uncertainty introduced by the inherent randomness in the model parameters and epistemic uncertainties due to lack of knowledge or imperfect modeling. Unfortunately, epistemic uncertainties are hard to quantify and cannot be completely eliminated for complex structures, though they can be reduced by using appropriate element types, better meshing and more sophisticated analysis techniques. Others have classified the modeling error by three components (e_1, e_2, e_3) summarized in Table 5-2, respectively representing the error due to discrepancy between the behavior of the mathematical model and that of the real structure, the error introduced in the numerical computation of the solution of partial differential equations, and the error arising from inaccurate assumptions made during simulation (Raphael and Smith 2003; Smith and Kripakaran 2008). The component e_3 may be further separated into two parts (e_{3a}, e_{3b}) which respectively arise from assumptions made when using the model (typically assumptions related to boundary conditions such as support characteristics and connection stiffness) and from errors in values of model parameters such as moment of inertia and Young's modulus. To compare with the notation used elsewhere (Liu and Frangopol 2004), aleatory uncertainties are represented by e_{3b} and epistemic uncertainties by e_1, e_2 and e_{3a}.

Additionally, measurement error (e_{meas}), the difference between real and measured quantities in a single measurement, is an important consideration that can affect the accuracy of structural identification (Shenton and Hu 2006a). Measurement errors result from equipment (precision, stability, robustness), as well as on-site installation faults (Sanayei et al. 1997). While it is tempting to quantify measurement error as a sum of individual sources, it is more reasonable to quantify them probabilistically using sensor precision and on-site information obtained during sensor installation.

Table 5-2. Types of uncertainty models

Model Error Classification	Type	Source
e_1	Epistemic	Discrepancy between the behavior of the mathematical model and real structure
e_2	Epistemic	Introduced by numerical computation of the solution
e_{3a}	Epistemic	Assumptions made when creating the model, i.e. boundary condition
e_{3b}	Aleatory	Error on values of model parameters, i.e. moment of inertia.

5.3.2 Objective Functions for Error Minimization

There are many physics based equations that are used as either objective functions or constraints for model updating, depending upon the parameters that are to be identified and the identification algorithm being used. No matter how they are cast, objective functions quantify the discrepancy between the model predictions and in-situ measurements or properties. A variety of measured responses and related parameters can be used for this purpose, including static displacements, tilts and strains (Sanayei and Saletnik 1996; Sanayei et al. 1997), boundary condition elements (Sanayei et al. 1999a), modal parameters (natural frequencies and mode shapes) (Moaveni et al. 2008; Sanayei et al. 1999a), modal strain energy (Lee and Yun 2006), higher order derivatives, and flexibility-based error functions (Arya et al. 1998; Farrar and Jauregui 1998; Hjelmstad and Shin 1997). There are also studies where static and modal data are combined (Oh and Jung 1998; Santini Bell et al. 2007; Wang et al. 2001).

For example, one may seek to define a threshold defined as the absolute value of the difference between measured responses (x_{meas}) and predicted responses (x_{mod}). The goal then is to calculate the RMS error of this residue over all the measurement points, identifying candidate models as those producing RMS errors below a specified threshold value. This is the fundamental strategy of the multiple model approach introduced earlier in this chapter. Once within the threshold, no distinction between candidate models can be made. Often the correct model has a high RMS value and wrong models have RMS values of zero. This formulation was further improved by combining errors using statistical methods (Ravindran et al. 2007). An example of this approach is given in Section 8.8.

The objective function can also be derived directly from the equations of motion to yield the "modal force error equation." This approach requires that mass and stiffness matrices from a linear elastic FEM be known prior to the identification process. The formulation of this objective function begins with the free-response, second-order equations of motion:

$$([K]-[M]\omega_i^2)\{\phi_i\} = \{0\} \tag{3}$$

Substituting the eigenvalues (modal frequencies) and eigenvectors (mode shapes) measured from the structure into this equation along with the mass and stiffness matrices from the original model yields a vector that is defined as the "modal force error," or "residual force." This vector represents the harmonic force vector that would have to be applied to the structure at this modal frequency in order to satisfy force equilibrium.

Optimization schemes for structural identification seek to minimize the objective function by variation of the parameters in the model. However, optimization schemes are driven by purely mathematical considerations. This fact can lead to the optimization scheme preferring to vary certain model parameters into physically inconsistent states. For example, perturbing a particular material density or a structural damping parameter to be negative may well minimize the objective function, but certainly the situations of negative inertia and negative damping (assuming there is no active control present) are not acceptable from a physical viewpoint. To avoid such situations, it is important to optimize the objective functions subject to constraints on the parameter values. These constraints will ensure that the parameters perturbed to minimize the objective function are kept within acceptable limits.

Constraints are typically enforced in one of two ways. The first way is "implicitly," in which the constraint is incorporated into the model form, such as the relationship between an element's elastic modulus and its elemental stiffness matrix. The second way a constraint can be enforced is "explicitly," in which the optimization scheme constrains the possible perturbations to the model parameters to

STRUCTURAL IDENTIFICATION OF CONSTRUCTED SYSTEMS 91

be within the limits imposed by the constraints. In this manner the possible perturbations to the parameters are reduced to a set that will still satisfy the constraints.

In modal parameter identification, constraints such as modal frequency and modal damping remaining positive are typically imposed explicitly after the optimization has been completed. Inability to find modal parameters that satisfy both the constraints and the optimization criteria is usually indicative of a poorly formed identification problem. Other constraints such as phase relationships between measurements and reciprocity between DOF are enforced implicitly in the form of the modal model.

In structural parameter identification, there are several constraints that are used either implicitly or explicitly, depending upon the optimization algorithm that is used. Preservation of the mass, damping, and stiffness matrix (referred to collectively as "property matrices") symmetry is used as a constraint. Preservation of the property matrix sparsity (the zero/nonzero pattern of the matrix) is also used as a constraint. The preservation of sparsity is one way to preserve the allowable load paths of the structure in the updated model. Preservation of the property matrix positivity is also used as a constraint. This constraint ensures that situations such as negative inertia or negative damping do not occur.

5.3.3 Model Calibration Techniques

The outcome of a successful St-ID application to an operating constructed system is an analytical model that simulates the mechanical characteristics of the structure and its behavior under realistic load effects and especially realistic intrinsic forces with an established level of confidence. Model calibration is commonly used, for example, to interpret measurement data from load tests on concrete bridges where values for flexural rigidity (i.e., the product of Young's modulus and the moment of inertia, EI) are determined (Burdet 1993). This strategy has also been used to find coefficients of stiffness matrices for vibration model calibration (Brownjohn et al. 2003; Doebling et al. 1998; Friswell and Mottershead 1995; Santini et al. 1999; Shmerling and Catbas 2009). This section (and a subsequent case study in Section 7.2 will now overview some of the approaches that have been employed to do so. It should be noted that model calibration techniques can also be used for the task of damage detection, where measured data from a structure with an unknown condition is compared to that of a reliable model of the undamaged structure (Hjelmstad and Shin 1997; Yao and Natke 1994). Therefore some of the methods presented reference applications to damage detection.

5.3.3.1 Manual, Heuristic-Based Model Calibration

One of the most basic approaches to model calibration is the parameter sensitivity study, to examine the impact of varying selected parameters on the simulated dynamic properties of a structure, and, thus, to determine the most critical parameters for the global model calibration. Parameter sensitivity based model updating can be conducted using an automated procedure or manually by analyzing the sensitivity of the parameters and understanding their impact on the structural response not only mathematically, but, also in physical terms. In such an approach, it is seen that that the most sensitive parameters can be material properties, boundary, and continuity conditions (Aktan et al. 1998). In bridges, critical properties include the elastic modulus of concrete, properties of rigid links used to connect shell and beam finite elements in decks, boundary conditions, and force releases and kinematics of the movement systems.

For the initial global calibration, the discrepancy between the global dynamic properties that are measured and the model responses are minimized by changing the identified sensitive parameters (Catbas et al. 2007). Models calibrated in this way are referred to as the "only globally calibrated" model of the structure. Thus, to permit local calibration of the numerical models, it is possible to conduct controlled load tests on the structure with known loads. In bridges, these may consist of a crawl speed test using a moving load and/or a static load pattern achieved by positioning trucks at various predetermined locations along the bridge while the sensors located at critical locations are used to collect measurement data from the sensor clusters. Local calibration can then be performed by comparing and correlating these experimental results with the corresponding FEM analytical responses. Once globally calibrated, the model's ability to simulate the modal properties can be investigated using an eigenvalue analysis of the model under different values of the most critical parameters of the structure, as applied to a long span bridge (Catbas et al. 2007).

5.3.3.2 Optimal Matrix Update Methods

One class of model updating is based on the modification of structural model matrices such as mass, stiffness, and damping to reproduce as closely as possible the measured static or dynamic response from the data. These methods solve for the updated matrices (or perturbations to the nominal model that produce the updated matrices) by forming a constrained optimization problem based on the structural equations of motion, the nominal model, and the measured data. Comparisons of the updated matrices to the original correlated matrices provide an indication of correlation of the model and the experimental data. Methods that use a closed-form direct solution to compute the updated structural matrices are commonly referred to as optimal matrix update methods (Hemez 1993; Kaouk 1993; Smith and Beattie 1991; Zimmerman and Smith 1992). The problem is generally formulated as a Lagrange multiplier or penalty-based optimization. Other studies (Baruch and Bar Itzhack 1978; Berman and Nagy 1983; Kabe 1985) have a common formulation of the optimal update problem that is essentially minimization of the Frobenius norm of global parameter matrix perturbations using zero modal force error and property matrix symmetry as constraints.

Smith and Beattie (1991) extend the formulation of (Kabe 1985) to include a sparsity preservation constraint (Smith 1992) and also formulate the problem as the minimization of both the perturbation matrix norm and the modal force error norm subject to the symmetry and sparsity constraints. Others (Brock 1968; Kammer 1988) similarly impose symmetry constraints in their minimization of the modal force error. Conversely, Chen and Garba (1988a; 1988b) present a method for minimizing the norm of the model property perturbations with a zero modal force error constraint. They also enforce a connectivity constraint to impose a known set of load paths onto the allowable perturbations. The updates are thus obtained at the element parameter level, rather than at the matrix level.

Another type of approach to the optimal matrix update problem involves the minimization of the rank of the perturbation matrix, rather than the norm of the perturbation matrix (Zimmerman and Kaouk 1994). Related to this, (Doebling 1996) presents a method to compute a minimum-rank update for the elemental parameter vector, rather than for global or elemental stiffness matrices. A limitation of this method as with all minimum-rank procedures is that the rank of the perturbation is always equal to the number of modes used in the computation of the modal force error.

Optimal update techniques have also been used for damage detection by formulating an overdetermined system for a set of damage parameters representing reductions in the extensional stiffness values for each member (Lindner et al. 1993). Lindner and Kirby (1994) later extended that technique to account for changes in elemental mass properties. Meanwhile, Liu (1995) presents an optimal update technique for computing the elemental stiffness and mass parameters for a truss structure from measured modal frequencies and mode shapes and use this method to locate a damaged truss element. Other global optimization techniques such as Coupled Local Minimizers have also been proposed (Bakir et al. 2008).

5.3.3.3 Sensitivity-Based Update Methods

Another class of matrix update methods, sensitivity-based update method, is based on the solution of a first-order Taylor series that minimizes an error function of the matrix perturbations (Hemez 1993). This involved the determination of a modified model parameter vector (consisting of material and/or geometric parameters), where the parameter perturbation vector is determined using the Newton-Raphson method. The main difference between the various sensitivity-based update schemes is the method used to estimate the sensitivity matrix, which can be either experimental or analytical quantities. For experimental sensitivity, the orthogonality relations can be used to compute the modal parameter derivatives. Analytical sensitivity methods usually require the evaluation of the stiffness and mass matrix derivatives, which are less sensitive than experimental sensitivity matrices to noise in the data and to large perturbations of the parameters.

Ricles (1991) presents a methodology for sensitivity-based matrix update, which takes into account variations in system mass and stiffness, center of mass locations, changes in natural frequency and mode shapes, and statistical confidence factors for the structural parameters and experimental instrumentation. The method uses a hybrid analytical/experimental sensitivity matrix, where the modal parameter sensitivities are computed from the experimental data, and the matrix sensitivities are computed from the analytical model. This method is further developed and applied to more numerical examples (Ricles and Kosmatka 1992).

Meanwhile, Sanayei and Onipede (1991) present a technique for updating the stiffness parameters of a FEM using the results of a static load-displacement test. A sensitivity-based, element-level parameter update scheme is used to minimize the residuals between the applied forces and forces produced by applying the measured displacements to the model stiffness matrix. In a related paper, Sanayei et al. (1992) explore this method's sensitivity to noise, and Sanayei and Saletnik (1996a; 1996b) then extend the algorithm and the error analysis to use static strain, rather than displacement, measurements. The sensitivity-based approach was expanded to include modal information (Sanayei et al. 1999b) and to use multi-response for simultaneous parameter estimation (Santini Bell et al. 2007).

5.3.3.4 Eigenstructure Assignment Method

Another matrix update method, known as "eigenstructure assignment," is based on the design of a fictitious controller whose gains are selected to minimize the modal force error (Lim 1994; Lim 1995). The controller gains are then interpreted as parameter matrix perturbations to the structural model. In addition to model calibration, this technique can be used to identify damage by comparing the "best achievable eigenvectors" yielded by the fictitious controller to the measured eigenvectors (Lim and Kashangaki 1994). Several researchers (Cobb and Liebst 1997a; Cobb and Liebst 1997a; Cobb and Liebst 1997b; Cobb and Liebst 1997b; Lindner and Goff 1993; Zimmerman and Kaouk 1992; Zimmerman and Kaouk 1992) also implemented such an eigenstructure assignment technique for damage detection. Lim (1994; 1995) later applied a constrained eigenstructure technique experimentally to a twenty-bay planar truss. This approach identifies element-level damage directly, rather than finding perturbations to the stiffness matrix. The computation of element-level perturbations is accomplished by diagonalizing the control gains, then interpreting the diagonal entries as changes to the elemental stiffness properties. The technique is shown to work well even with limited instrumentation. Finally, Schulz et al. (1996) present a technique similar to eigenstructure assignment known as "FRF assignment." The authors formulate the problem as a linear solution for element-level stiffness and mass perturbation factors, directly using the FRFs instead of requiring the added step of mode shape extraction.

5.3.3.5 Hybrid Matrix Update Methods

Kim and Bartkowicz (1993) and Kim et al. (1995) present a two-step model calibration technique used for damage-detection in large structures with limited instrumentation. The first step uses optimal matrix update to identify the region of the structure where damage has occurred. The second step is a sensitivity-based method, which locates the specific structural element where damage has occurred. Li and Smith (1994; 1995) and Sanayei et al. (2006) also present a hybrid model updating technique that constrains the stiffness matrix perturbation to preserve the connectivity of the FEM with greater computational efficiency than the iterative sparsity-preserving algorithm (Smith 1992).

5.3.3.6 Evolutionary Optimization

Evolutionary optimization is part of a larger mathematical field called stochastic search. Evolutionary Algorithms (EAs) are well suited to global optimization (Bèack 1996). These methods exploit a set of potential solutions referred to as "population" and approach the global optimum through cooperation and competition among the individuals in the population to avoid the shortcomings of classical optimization methods (Franco et al. 2004). Another advantage of evolutionary optimization is that such algorithms do not require continuity and differentiability of objective functions, and hence, construction of the objective functions is usually more flexible than other approaches, e.g., can be formulated in terms of frequency, mode shape or even modal assurance criteria (MAC). Among the various paradigms of EAs, genetic algorithms (GA) (Chou and Ghaboussi 2001; Goldberg 1989) have been most frequently employed for model parameter identification and damage detection (Friswell et al. 1998; Mares and Surace 1996; Shenton and Hu 2006b). These approaches have been successfully employed to identify civil structures under various excitation conditions (Boller and Staszewski 2004; Casciati 2003; Chang 2005; Housner et al. 1997; Johnson and Smyth 2006; S. C. Liu 2003; Lu and Tu 2005; Natke and Yao 1988; Natke et al. 1993; Smyth and Betti 2004).

With EAs the implementation of multi-objective functions in the context of FEM updating becomes rather straightforward. Performing simultaneous optimization with several objective functions can avoid the difficulty arising from the need to choose appropriate weights to combine different types of measured data into a single objective function. To improve the computational efficiency, the parameter identification of a large system can be carried out in a staged procedure, such that the occurrence and approximate location of the parameter changes are identified first by a coarse model, whereas the precise location and the extent of damage are determined in a subsequent stage using a model involving a refined discretization for the damaged region. Perera and Ruiz (2008) carried out a multi-stage FEM updating procedure based on a multi-objective evolutionary optimization. The multi-objective optimization problem is solved by the so-called Strength Pareto Genetic Algorithm (SPGA), which implements a fitness sharing scheme combined with elitism.

5.4 Closing Remarks

Today, monitoring efforts worldwide continue the task of determining in-situ behavior with the added emphasis on rapid assessment and evaluation. Unfortunately, this has not fully translated into practice, as full-scale evaluation and model identification have not become standard practice for civil infrastructure management. Reasons include difficulties and initial costs associated with testing large structural systems and the inability to safely test the structures at the ultimate limit states of most concern, since tests that can be used to verify analytical predictions of the ultimate load capacity would, by definition, be damaging to the structure. Another challenge to incorporation of structural identification into the assessment of the performance of constructed facilities is the weaknesses of model creation during the initial design (Santini-Bell 2008). Full-scale observations provide information essential to advancing the design state-of-the-art, including:

- Quantitative knowledge on the as-built state parameters (e.g., initial stresses, strains and displacements, local and global stiffness) and their variation at different limit-states and over time;

- Quantitative definitions for performance parameters (e.g., functionality, serviceability, safety, life-cycle cost, etc.) and relationships between the state and performance parameters;

- Understanding of the loading environment (including intrinsic and environmental loads) and defects, deterioration and damage mechanisms which influence the state-of-force in a structure, lead to changes in state parameters, and/or affect performance;
- Real capacities of load-resisting mechanisms and how the capacities and failure mechanisms are affected by various types of defects, deterioration, and damage;
- Observable and easily measurable condition-and-damage indices sensitive to changes in state properties, and that relate directly to the capacity of load resisting mechanisms and performance.

Nevertheless, structural identification needs to be advanced in areas such as:

- Uncertainty quantification and correlation
- Application and adaptation of advanced computing and mathematical methods, including data mining, feature selection, stochastic search and signal analysis
- Sensor network data clustering and interpretation
- Model sampling in combinatorial situations
- Integration of non-physics-based data interpretation and physics-based data interpretation within an integrated framework for condition assessment
- Identification reliability to avoid costly false positives and false negatives
- Effect of measurement on structural reliability
- Navigating spaces of candidate models
- Improved evaluation of errors in models, including modeling assumptions and model parameters
- As-installed measurement errors both from sensors and data acquisition
- Field studies that demonstrate quantitative and qualitative benefits to practitioners
- Combined data interpretation using static and dynamic measurements
- Integration of structural identification into the initial design process.

5.5 References

Aktan, A. E., Farhey, D. N., Helmicki, A. J., Brown, D. L., Hunt, V. J., Lee, K. L., and Levi, A. (1997). "Structural identification for condition assessment: Experimental arts." *Journal of Structural Engineering,* 123(12), 1674-1684.

Aktan, A. E., Helmicki, A. J., and Hunt, V. J. (1998). "Issues in health-monitoring for intelligent infrastructure." *Smart Materials and Structures,* 7(5), 674-692.

Alimoradi, A., Miranda, E., Taghavi, S., and Naeim, F. (2006). "Evolutionary modal identification utilizing coupled shear-flexural response-implication for multistory buildings. part I: Theory." *The Structural Design of Tall Buildings*, 15, 51-65.

Alimoradi, A., and Naeim, F. (2006). "Evolutionary modal identification utilizing coupled shear-flexural response-implication for multistory buildings. part II: Application." *The Structural Design of Tall Buildings*, 15, 67-103.

Allemang, R. J. (1995). *Vibrations: Experimental modal analysis,* UC-SDRL-CN-20-263-663/664, University of Cincinnati, Cincinnati, OH.

Allemang, R. J., Brown, D. L., and Fladung, W. (1994). "Modal parameter estimation: A unified matrix polynomial approach." *12th International Modal Analysis Conference,* Society of Experimental Mechanics, Honolulu, Hawaii.

Areemit, N., Yamaguchi, H., Matsumoto, Y., and Ibi, T. (2003). "Model identification of a four story reinforced concrete building under renovation using ambient vibration measurement." *Proceedings of Structural Health Monitoring and Intelligent Infrastructure,* Tokyo, Japan.

Arya, B., Santini, E. M., and Sanayei, M. (1998). "Impact of finite element modeling error in model updating." *2nd World Conference on Structural Control,* John Wiley and Sons,

Bakir, P. G., Reynders, E., and Roeck, G. (2008). "An improved finite element model updating method by the global optimization technique 'coupled local minimizers'." *Computers and Structures,* 86(11), 1339-1352.

Banan, M. R., and Hjelmstad, K. D. (1994). "Parameter estimation of structures from static response. II: Numerical simulation studies." *Journal of Structural Engineering,* 120(11), 3259-3283.

Baruch, M., and Bar Itzhack, I. Y. (1978). "Optimum weighted orthogonalization of measured modes." *AIAA Journal,* 16(4), 346-351.

Bèack, T. (1996). *Evolutionary algorithms in theory and practice: Evolution strategies, evolutionary programming, genetic algorithms,* Oxford University Press, New York.

Beck, J. L. (1978). *Determining models of structures from earthquake records,* EERL Report No. 78-01, California Institute of Technology, Pasadena, CA.

Beck, J. L., and Jennings, P. (1980). "Structural identification using linear models and earthquake records." *Earthquake Engineering and Structural Dynamics,* 8, 145-160.

Berman, A., and Nagy, E. J. (1983). "Improvement of large analytical model using test data." *AIAA Journal*, 21(8), 1168-1173.

Bernal, D. (2006). "Flexibility-based damage localization from stochastic realization results." *Journal of Engineering Mechanics*, 132(6), 651-658.

Bernal, D., and Gunes, B. (2004). "Flexibility based approach for damage characterization: Benchmark application." *Journal of Engineering Mechanics*, 130, 61-70.

Boller, C., and Staszewski, W. J., Eds. (2004). *Proceedings of the second european workshop on structural health monitoring*, DEStech Publications, Inc., Lancaster, PA.

Brincker, R., Zhang, L., and Andersen, P. (2001). "Modal identification of output-only systems using frequency domain decomposition." *Smart Materials and Structures*, 10(3), 441-445.

Brock, J. E. (1968). "Optimal matrices describing linear systems." *AIAA Journal*, 6(7), 1292-1296.

Brown, D. L., Allemang, R. J., Zimmerman, R., and Mergeay, M. (1979). *Parameter estimation techniques for modal analysis*, 790221, SAE International, Warrendale, PA.

Brownjohn, J. M. W., Moyo, P., Omenzetter, P., and Lu, Y. (2003). "Assessment of highway bridge upgrading by dynamic testing and finite element model updating." *ASCE Journal of Bridge Engineering*, 8(3), 162-172.

Burdet, O. (1993). "Load testing and monitoring of swiss bridges." *Safety and Performance Concepts, Bulletin 219*, Lausanne, Switzerland.

Caicedo, J. M., Dyke, S. J., and Johnson, E. A. (2004). "Natural excitation technique and eigensystem realization algorithm for phase i of the IASC-ASCE benchmark problem: Simulated data." *ASCE Journal of Engineering Mechanics*, 130(1), 49-60.

Campbell, S., Kwok, K. C. S., and Hitchcock, P. A. (2005). "Dynamic characteristics and wind-induced response of two high-rise residential buildings during typhoons." *J.Wind Eng.Ind.Aerodyn.*, 93, 461.

Casciati, F., Ed. (2003). *Proceedings of third world conference on structural control*, John Wiley and Sons, Chichester, West Sussex, UK.

Catbas, F. N., Gul, M., and Burkett, J. (2008). "Conceptual damage-sensitive features for structural health monitoring: Laboratory and field demonstrations." *Mechanical Systems and Signal Processing*, 22(7), 1650-1669.

Catbas, F. N., Brown, D. L., and Aktan, A. E. (2004). "Parameter estimation for multiple-input multiple-output modal analysis of large structures." *Journal of Engineering Mechanics*, 130(8), 921-930.

Catbas, F. N., Ciloglu, S. K., Hasancebi, O., Grimmelsman, K. A., and Aktan, A. E. (2007). "Limitations in structural identification of large constructed structures." *Journal of Structural Engineering*, 133(8), 1051-1066.

Celebi, M. (1993). "Seismic response of eccentrically braced tall building." *Journal of Structural Engineering*, 119(4), 1188-1205.

Celebi, M. (1996). "Comparison of damping in buildings under low-amplitude and strong motions." *Journal of Wind Engineering and Industrial Aerodynamics*, 59(2-3), 309-323.

Celebi, M. (2006). "Recorded earthquake responses from the integrated seismic monitoring network of the Atwood building, anchorage, alaska." *Earthquake Spectra*, 22(4), 847-864.

Chang, F. K., Ed. (2005). *Proceedings of the fifth international workshop on structural health monitoring*, Stanford University, Palo Alto, CA.

Chen, J. C., and Garba, J. A. (1988a). "On-orbit damage assessment for large space structures." *AIAA Journal*, 26(9), 1119-1126.

Chen, J. C., and Garba, J. A. (1988b). "Structural damage assessment using a system identification technique." *Structural safety evaluation based on system identification approaches*, H. G. Natke and J. T. P. Yao, eds., Friedr. Vieweg & Sohn, Braunschweig, Wiesbaden, Germany, 1.

Ching, J., Beck, J. L., and Porter, K. A. (2006). "Bayesian state and parameter estimation of dynamical systems." *Probabilistic Engineering Mechanics*, 21(1), 81-96.

Chou, J. H., and Ghaboussi, J. (2001). "Genetic algorithm in structural damage detection." *Computers and Structures*, 79, 1335-1353.

Cobb, R. G., and Liebst, B. S. (1997a). "Sensor placement and structural damage identification from minimal sensor information." *AIAA Journal*, 35(2), 369-374.

Cobb, R. G., and Liebst, B. S. (1997b). "Structural damage identification using assigned partial eigenstructure." *AIAA Journal*, 35(1), 152-158.

Cole, H. A. (1973). *On-line failure detection and damping measurement of aerospace structures by random decrement signatures*, NASA CR-2205, NASA, Washington, DC.

Davenport, A. G., and Hill-Carroll, P. (1986). "Damping in tall buildings: Its variability and treatment in design." *Building motion in wind,* N. Isyumov and T. Tschanz, eds., ASCE, New York, 42-57.

Deblauwe, F., Allemang, R. J., and Allemang, R. J. (1985). "The polyreference time-domain technique." *10th International Seminar on Modal Analysis, Part IV,* Katholieke Universiteit, Leuven, Belgium.

Doebling, S. W., Farrar, C. R., and Prime, M. B. (1998). "A summary review of vibration-based damage identification methods." *Shock and Vibration Digest,* 30(2), 91-105.

Doebling, S. W. (1996). "Minimum-rank optimal update of elemental stiffness parameters for structural damage identification." *AIAA Journal,* 34(12), 2615-2621.

Duan, Z., Yan, G., Ou, J., and Spencer, B. F. (2005). "Damage localization in ambient vibration by constructing proportional flexibility matrix." *Journal of Sound and Vibration,* 284, 455-466.

Durrani, A. J., Mau, S. T., AbouHashish, A. A., and Li, Y. (1994). "Earthquake response of flat-slab buildings." *Journal of Structural Engineering,* 120, 947-964.

Erwin, S., Kijewski-Correa, T., and Yoon, S. Y. (2007). "Full-scale verification of dynamic properties from short duration records." *Structures Congress 2007,* Long Beach, CA.

Fang, J. Q., Li, Q. S., Jeary, A. P., and Liu, D. K. (1999). "Damping of tall buildings: Its evaluation and probabilistic characteristics." *The Structural Design of Tall Buildings,* 8, 145.

Farrar, C. R., and Jauregui, D. A. (1998). "Comparative study of damage identification algorithms applied to a bridge: I. experiment." *Smart Materials and Structuresand,* 7(5), 704-719.

Farrar, C. R., Sohn, H., Hemez, F. M., Anderson, M. C., Bement, M. T., Corwell, P. J., Doebling, S. W., Schultze, J. F., Lieven, N., and Robertson, A. N. (2003). *Damage prognosis: Current status and future needs,* LA-14051-MS, Los Alamos National Laboratory, Los Alamos, NM.

Franco, G., Betti, R., and Lus, H. (2004). "Identification of structural systems using an evolutionary strategy." *ASCE Journal of Engineering Mechanics,* 130(10), 1125-1139.

Frangopol, D. M., and Liu, M. (2007). "Maintenance and management of civil infrastructure based on condition, safety, optimization." *Structure and Infrastructure*

Engineering: Maintenance, Management, Life-Cyle Design and Performance, 3, 29-41.

Friswell, M. I., and Mottershead, J. E. (1995). *Finite element modal updating in structural dynamics,* Kluwer Academic Publishers, Dordrecht, The Netherlands.

Friswell, M. I., Penny, J. E. T., and Garvey, S. D. (1998). "A combined genetic and eigensensitivity algorithm for the location of damage in structures." *Computers and Structures,* 69(8), 547-556.

Ghanem, R., and Shinozuka, M. (1995). "Structural-system identification. I: Theory." *Journal of Engineering Mechanics,* 121(2), 255-264.

Goldberg, D. E. (1989). *Genetic algorithms in search, optimization and machine learning,* Addison-Wesley Longman, Reading, MA.

Guo, S. J. (2002). "Improvement of a tail-plane structural model using vibration test data." *Journal of Sound and Vibration,* 256(4), 647-663.

Gurley, K., and Kareem, A. (1999). "Applications of wavelet transform in earthquake, wind and ocean engineering." *Engineering Structures,* 21(2), 149-167.

Hashimoto, P. S., Steele, L. K., Johnson, J. J., and Mensing, R. W. (1993). *Review of structural damping values for elastic seismic analysis of nuclear power plants,* NUREG/CR-6011, United States Nuclear Regulatory Commission, Washington, DC.

Hemez, F. M. (1993). "Theoretical and experimental correlation between finite element models and modal tests in the context of large flexible space structures." PhD Dissertation, University of Colorado, Boulder, CO.

Hjelmstad, K. D., and Shin, S. (1997). "Damage detection and assessment of structures from static responses." *ASCE Journal of Engineering Mechanics,* 123(6), 568-576.

Hoshiya, M., and Saito, E. (1984). "Structural identification by extended kalman filter." *Journal of Engineering Mechanics,* 110(12), 1757-1772.

Housner, G. W., Bergman, L. A., Caughey, T. K., Chassiakos, A. G., Claus, R. O., Masri, S. F., Skelton, R. E., Soong, T. T., Spencer, B. F., and Yao, J. T. P. (1997). "Structural control: Past, present, and future." *Journal of Engineering Mechanics,* 123(9), 897-971.

James, G. H., Carne, T. G., and Lauffer, J. P. (1993). *The natural excitation technique for modal parameter extraction from operating wind turbines,* SAND92-1666, UC-261, Sandia National Laboratories, Albuquerque, NM.

Jang, J. H., Yeo, I., Shin, S., and Chang, S. P. (2002). "Experimental investigation of system-identification-based damage assessment on structures." *ASCE Journal of Structural Engineering*, 128(5), 673-682.

Jeary, A. P. (1992). "Establishing nonlinear damping characteristics from non-stationary response time histories." *The Structural Engineer*, 70(4), 61-66.

Johnson, E. A., and Smyth, A. W., Eds. (2006). *Proceedings of the 4th world conference on structural control and monitoring (4WCSCM)*, International Association for Structural Control and Monitoring/USC, Los Angeles, CA.

Juang, J. N. (1994). *Applied system identification*, PTR Prentice Hall, Englewood Cliffs, NJ.

Juang, J. N., and Pappa, R. S. (1985). "An eigensystem realization algorithm for modal parameter identification and model reduction." *AIAA Journal of Guidance, Control, and Dynamics*, 8(5), 620-627.

Kabe, A. M. (1985). "Stiffness matrix adjustment using mode data." *AIAA Journal*, 23(9), 1431-1436.

Kammer, D. C. (1988). "Optimal approximation for residual stiffness in linear system identification." *AIAA Journal*, 26(1), 104-112.

Kaouk, M. (1993). "Finite element model adjustment and damage detection using measured test data." PhD Dissertation, University of Florida, Gainesville, FL.

Kijewski-Correa, T., Kilpatrick, J., Kareem, A., Kwon, D. K., Bashor, B., Kochly, M., Young, B. S., Abdelrazaq, A., Galsworthy, J. K., Isyumov, N., Morrish, D., Sinn, R. C., and Baker, W. F. (2006). "Validating wind-induced response of tall buildings: Synopsis of the Chicago full-scale monitoring program." *ASCE Journal of Structural Engineering*, 132(10), 1509-1523.

Kijewski-Correa, T., and Pirnia, J. D. (2007). "Dynamic behavior of tall buildings under wind: Insights from full-scale monitoring." *Structural Design of Tall and Special Buildings*, 16, 471-486.

Kijewski-Correa, T., and Cycon, J. (2007). "System identification of constructed buildings: Current state-of-the-art and future directions." *Structural Health Monitoring and Intelligent Infrastructure*, Vancouver, Canada.

Kijewski-Correa, T., Taciroglu, E., and Beck, J. L. (2008). "System identification of constructed facilities: Challenges and opportunities across hazards." *Structures Congress*, Vancouver, Canada.

Kim, H. M., and Bartkowicz, T. J. (1993). "Damage detection and health monitoring of large space structures." *Sound and Vibration*, 27(6), 12-17.

Kim, H. M., Bartkowicz, T. J., Smith, S. W., and Zimmerman, D. C. (1995). "Structural health monitoring of large structures." *49th Meeting of the Society for Machinery Failure Prevention Technology*,

Kong, J. S., and Frangopol, D. M. (2003). "Life-cycle reliability-based maintenance cost optimization of deteriorating structures with emphasis on bridges." *Journal of Structural Engineering*, 129(6), 818-828.

Lagomarsino, S., and Pagnini, L. C. (1995). *Criteria for modeling and predicting dynamic parameters of buildings*, ISC-II, Instituto di Scienza Delle Costruzioni, University of Genova, Genova, Italy.

Lee, J. J., and Yun, C. B. (2006). "Two-step approaches for effective bridge health monitoring." *Structural Engineering and Mechanics*, 23(1), 75-95.

Li, S. J., Suzuki, Y., and Noori, M. (2004). "Identification of hysteretic systems with slip using bootstrap filter." *Mechanical Systems and Signal Processing*, 18(4), 781-795.

Li, Y., and Mau, S. (1991). "A case study of MIMO system identification applied to building seismic records." *Earthquake Engineering and Structural Dynamics*, 20, 1045-1064.

Li, C., and Smith, S. W. (1994). "A hybrid approach for damage detection in flexible structures." *35th AIAA/ASME/ASCE/AHS/ASC Structures, Structural Dynamics, and Materials Conference*,

Li, C., and Smith, S. W. (1995). "Hybrid approach for damage detection in flexible structures." *Journal of Guidance, Control, and Dynamics*, 18(3), 419-425.

Li, Q. S., Fu, J. Y., Xiao, Y. Q., Li, Z. N., and Ni, Z. H. (2006). "Wind tunnel and full-scale study of wind effects on China's tallest building." *Eng.Struct.*, 28, 1745-1758.

Li, Q. S., Wu, J. R., Liang, S. G., Xiao, Y. Q., and Wong, C. K. (2004). "Full-scale measurements and numerical evaluation of wind-induced vibration of a 63-story reinforced concrete tall building." *Eng.Struct.*, 26, 1779-1794.

Li, Q. S., Xiao, Y. Q., and Wong, C. K. (2005). "Full-scale monitoring of typhoon effects on super tall buildings." *J.Fluids Struct.*, 20, 697-717.

Li, Q. S., Xiao, Y. Q., Wong, C. K., and Jeary, A. P. (2004). "Field measurements of typhoon effects on a super tall building." *Eng.Struct.*, 26, 233-244.

Li, Q. S., Yang, K., Wong, C. K., and Jeary, A. P. (2003). "The effect of amplitude-dependent damping on wind-induced vibrations of a super tall building." *J.Wind Eng.Ind.Aerodyn.*, 91, 1175-1198.

Lim, T. W. (1994). "Structural damage detection of a planar truss structure using a constrained eigenstructure assignment." *35th AIAA/ASME/ ASCE/AHS/ASC Structures, Structural Dynamics and Materials Conference*, Hilton Head, SC.

Lim, T. W. (1995). "Structural damage detection using constrained eigenstructure assignment." *Journal of Guidance, Control, and Dynamics*, 18(3), 411-418.

Lim, T. W., and Kashangaki, T. A. L. (1994). "Structural damage detection of space truss structure using best achievable eigenvectors." *AIAA Journal*, 32(5), 1049-1057.

Lin, B. C., and Papageorgiou, A. S. (1989). "Demonstration of torsional coupling caused by closely spaced periods: 1984 Morgan Hill earthquake response of the Santa Clara county building." *Earthquake Spectra*, 5(3), 539-556.

Lindner, D. K., and Goff, R. (1993). "Damage detection, location and estimation for space trusses." *Proceedings vol. 1917, smart structures and materials 1993: Smart structures and intelligent systems*, SPIE, Bellingham, WA, 1028-1039.

Lindner, D. K., and Kirby, G. (1994). "Location and estimation of damage in a beam using identification algorithms." *35th AIAA/ASME/ ASCE/AHS/ASC Structures, Structural Dynamics and Materials Conf.* Hilton Head, SC.

Lindner, D. K., Twitty, G. B., and Osterman, S. (1993). "Damage detection for composite materials using dynamic response data." *ASME Adaptive Structures and Materials Systems*, AD 35, 441-448.

Liu, M., and Frangopol, D. M. (2004). "Optimal bridge maintenance planning based on probabilistic performance prediction." *Engineering Structures*, 26(7), 991-1002.

Liu, S. C., Ed. (2003). *Proceedings of smart systems and NDE for civil infrastructures*, SPIE, San Diego, CA.

Liu, P. L. (1995). "Identification and damage detection of trusses using modal data." *Journal of Structural Engineering*, 121(4), 599-608.

Ljung, L. (1998). *System identification - theory for the user*, Prentice Hall, Englewood Cliffs, NJ.

Loh, C., and Tou, I. (1995). "A system identification approach to the detection of changes in both linear and non-linear structural parameters." *Earthquake Engineering and Structural Dynamics*, 24, 85-97.

Lu, Y., and Tu, Z. (2005). "Dynamic model updating using combined genetic-eigensensitivity algorithm and application in seismic response prediction." *Earthquake Engineering and Structural Dynamics*, 34(9), 1149-1170.

Maia, N. M. M., and Silva, J. M. M., Eds. (1997). *Theoretical and experimental modal analysis*, John Wiley and Sons, Inc., New York, NY.

Mares, C., and Surace, C. (1996). "An application of genetic algorithms to identify damage in elastic structures." *Journal of Sound Vibration*, 195(2), 195-215.

Masri, S. F., Miller, R. K., Saud, A. F., and Caughey, T. K. (1987). "Identification of nonlinear vibrating structures: Part I - formulation." *ASME Journal of Applied Mechanics*, 54, 918-922.

Mau, S. T., and Aruna, V. (1994). "Story drift, shear and OTM estimation from building seismic records." *Journal of Structural Engineering*, 120, 3366-3385.

McVerry, G. H. (1980). "Structural identification in the frequency domain from earthquake records." *Earthquake Engineering and Structural Dynamics*, 8, 161-180.

McVerry, G. H., and Beck, J. L. (1983). *Structural identification of JPL building 180 using optimally synchronized earthquake records*, EERL Report No. 83-01, California Institute of Technology, Pasadena, CA.

Mendel, J. M. (1995). *Lessons in estimation theory for signal processing, communications, and control*, Prentice Hall, New York, NY.

Mita, A., and Hagiwara, H. (2003). "Damage diagnosis of a building structure using support vector machine and modal frequency patterns." *Proceedings of Smart Systems and Nondestructive Evaluation for Civil Infrastructures*, SPIE, San Diego, CA.

Moaveni, B., He, X., Conte, J. P., and de Callafon, R. A. (2008). "Damage identification of a composite beam using finite element model updating." *Journal of Computer Aided Civil and Infrastructure Engineering*, 23(5), 339-359.

Nagarajaiah, S., and Li, Z. (2004). "Time segmented least squares identification of base isolated buildings." *Soil Dynamics and Earthquake Engineering*, 24, 577-586.

Natke, H. G., Tomlinson, G. R., and Yao, J. T. P., Eds. (1993). *Proceedings of the international workshop on safety evaluation based on identification approaches related to time variant and nonlinear structures*, Friedrich Vieweg and Sohn Verlag, Lambrecht, Germany.

Natke, H. G., and Yao, J. T. P., Eds. (1988). *Structural safety evaluation based on system identification approaches,* Friedrich Vieweg and Sohn Verlag, Braunschweig, Germany.

Nisar, A., Werner, S. D., and Beck, J. L. (1992). "Assessment of UBC seismic design provisions using recorded building motions." *10th World Conference on Earthquake Engineering,* Madrid, Spain.

Oh, B. H., and Jung, B. S. (1998). "Structural damage assessment with combined data of static and model tests." *ASCE Journal of Structural Engineering,* 124(8), 956-965.

Oh, C. K., and Beck, J. L. (2006). "Sparse Bayesian learning for structural health monitoring." *4th World Conference on Structural Control and Monitoring,* San Diego, CA.

Papageorgiou, A. S., and Lin, B. C. (1989a). "Influence of lateral-load-resisting system on the earthquake response of structures – a system identification study." *Earthquake Engineering and Structural Dynamics,* 18, 799-814.

Papageorgiou, A. S., and Lin, B. C. (1989b). "Study of the earthquake response of the base-isolated law and justice center in Rancho Cucamonga." *Earthquake Engineering and Structural Dynamics,* 18, 1189-1200.

Papageorgiou, A. S., and Lin, B. C. (1991a). "Analysis of recorded earthquake response identification of a multi story structure accounting for foundation interaction effects." *Soil Dynamics and Earthquake Engineering,* 9, 55-64.

Papageorgiou, A. S., and Lin, B. C. (1991b). "Earthquake response of two repaired buildings in past seismic shaking." *Soil Dynamics and Earthquake Engineering,* 10(5), 236-248.

Pavic, A., Brownjohn, J. M. W., and Živanović, S. (2010). "VSATs software for assessing and visualising floor vibration serviceability based on first principles." *Structures Congress,* ASCE, Orlando, FL.

Perera, R., and Ruiz, A. (2008). "A multistage FE updating procedure for damage identification in large-scale structures based on multi-objective evolutionary optimization." *Mechanical Systems and Signal Processing,* 22, 970-991.

Pirnia, J. D., Kijewski-Correa, T., Abdelrazaq, A., Chung, J., and Kareem, A. (2007). "Full-scale validation of wind-induced response of tall buildings: Investigation of amplitude dependent properties." *Structures Congress 2007,* Long Beach, CA.

Rainieri, C., Fabbrocino, G., Manfredi, G., and Cosenza, E. (2007). "Structural monitoring and assessment of the school of engineering main building at University of Naples Federico II." *3rd International Conference on Structural Health Monitoring and Intelligent Infrastructure*, Vancouver, BC.

Raphael, B., and Smith, I. F. C. (2003). *Fundamentals of computer aided engineering*, John Wiley and Sons, Chichester, UK.

Ravindran, S., Kripakaran, P., and Smith, I. F. C. (2007). "Evaluating reliability of multiple-model system identification." *24th W78 Conference*, Maribor, Slovenia.

Ricles, J. M. (1991). *Nondestructive structural damage detection in flexible space structures using vibration characterization*, CR-185670, NASA, Washington, DC.

Ricles, J. M., and Kosmatka, J. B. (1992). "Damage detection in elastic structures using vibratory residual forces and weighted sensitivity." *AIAA Journal*, 30, 2310-2316.

Robert-Nicoud, Y., Raphael, B., Burdet, O., and Smith, I. F. C. (2005). "Model identification of bridges using measurement data." *Computer-Aided Civil and Infrastructure Engineering*, 20(2), 118-131.

Robert-Nicoud, Y., Raphael, B., and Smith, I. F. C. (2005). "System identification through model composition and stochastic search." *Journal of Computing in Civil Engineering*, 19(3), 239-247.

Rodgers, J., and Celebi, M. (2006). "Seismic response and damage detection analyses of an instrumented steel moment-framed building." *Journal of Structural Engineering*, 132(10), 1543-1552.

Safak, E. (1989a). "Adaptive modeling, identification, and control of dynamic structural systems. I: Theory." *Journal of Engineering Mechanics*, 115(11), 2386-2405.

Safak, E. (1989b). "Adaptive modeling, identification, and control of dynamic structural systems. II: Applications." *Journal of Engineering Mechanics*, 115(11), 2406-2425.

Safak, E. (1991). "Identification of linear structures using discrete-time filters." *Journal of Structural Engineering*, 117(10), 3064-3085.

Safak, E. (1993). "Response of a 42-story steel-frame building to the Ms=7.1 Loma Prieta Earthquake." *Engineering Structures*, 15, 403-421.

Safak, E., and Celebi, M. (1991). "Seismic response of Transamerica Building: Part II system identification." *Journal of Structural Engineering*, 117, 2405-2425.

Safak, E., and Celebi, M. (1992). "Recorded seismic response of Pacific Park Plaza: Part II- system identification." *Journal of Structural Engineering,* 118, 1566-1589.

Saito, T., and Yokota, H. (1996). "Evaluation of dynamic characteristics of high-rise buildings using system identification techniques." *Journal of Wind Engineering and Industrial Aerodynamics,* 59, 299-307.

Saitta, S., Kripakaran, P., Raphael, B., and Smith, I. F. C. (2008). "Improving system identification using clustering." *Journal of Computing in Civil Engineering,* 22(5), 292-302.

Sanayei, M., Bell, E. S., Jvdekar, C. J., Edelmann, J. L., and Slavsky, E. (2006). "Damage localization and finite element model updating using multi-response NDT data." *Journal of Bridge Engineering,* (12), 689-699.

Sanayei, M., Imbaro, G. R., McClain, J. A. S., and Brown, L. C. (1997). "Structural model updating using experimental static measurements." *Journal of Structural Engineering,* 123(6), 792-798.

Sanayei, M., McClain, J. A. S., Wadia-Fascetti, S., and Santini, E. M. (1999a). "Parameter estimation incorporating modal data and boundary conditions." *Journal of Structural Engineering,* 125(9), 1048-1055.

Sanayei, M., McClain, J. A. S., Wadia-Fascetti, S., and Santini, E. M. (1999b). "Parameter estimation incorporating modal data and boundary conditions." *Journal of Structural Engineering,* 125(9), 1048-1055.

Sanayei, M., and Saletnik, M. J. (1996). "Parameter estimation of structures from static strain measurements." *Journal of Structural Engineering,* 122(5), 555-562.

Sanayei, M., Santini-Bell, E., Javdekar, C. N., and Edelmann, J. L. (2006). "Damage localization and finite element model updating using multi-response NDT data." *Journal of Bridge Engineering,* 11(6), 688-698.

Sanayei, M., and Onipede, O. (1991). "Damage assessment of structures using static test data." *AIAA Journal,* 29(7), 1174-1179.

Sanayei, M., Onipede, O., and Babu, S. R. (1992). "Selection of noisy measurement locations for error reduction in static parameter identification." *AIAA Journal,* 30(9), 2299-2309.

Sanayei, M., and Saletnik, M. J. (1996a). "Parameter estimation of structures from static strain measurements, part I: Formulation." *Journal of Structural Engineering-ASCE,* 122(5), 555-562.

Sanayei, M., and Saletnik, M. J. (1996b). "Parameter estimation of structures from static strain measurements, part II: Error sensitivity analysis." *Journal of Structural Engineering-ASCE,* 122(5), 563-572.

Santini Bell, E., Sanayei, M., Javdekar, C. N., and Slavsky, E. (2007). "Multiresponse parameter estimation for finite-element model updating using nondestructive test data." *Journal of Structural Engineering,* 133(8), 1067-1079.

Santini, E. M., Sanayei, M., Liu, M., and Olson, L. D. (1999). "Bridge foundation stiffness identification using dynamic field measurements." *Structures Congress,* New Orleans, LA.

Santini-Bell, E. (2008). "A bridge condition assessment framework integrating structural health monitoring and intelligent transportation technology." *Proceedings of the 2008 Transportation Research Board Annual Meeting,* Washington, DC.

Satake, N., Suda, K., Arakawa, T., Sasaki, A., and Tamura, Y. (2003). "Damping evaluation using full-scale data of buildings in Japan." *J.Struct.Eng.,* 129(4), 470-477.

Sato, T., and Qi, K. (1998). "Adaptive H1 filter: Its application to structural identification." *Journal of Engineering Mechanics,* 124(11), 1233-1240.

Sato, T., and Qi, K. (1999). "Nonlinear structural identification based on H1 filter algorithm." *SPIE,* 3667, 394-405.

Schulz, M. J., Pai, P. F., and Abdelnaser, A. S. (1996). "Frequency response function assignment technique for structural damage identification." *14th International Modal Analysis Conference,* Dearborn, MI.

Shenton, H. A., and Hu, X. (2006a). "Damage identification based on dead load redistribution: Effect of measurement error." *Journal of Structural Engineering,* 132(8), 1264-1273.

Shenton, H. A., and Hu, X. (2006b). "Damage identification based on dead load redistribution: Methodology." *Journal of Structural Engineering,* 132(8), 1254-1263.

Shmerling, R. Z., and Catbas, F. N. (2009). "Load rating and reliability analysis of an aerial guideways." *Journal of Bridge Engineering,* 14(4), 247-256.

Skolnik, D., Lei, Y., Yu, E., and Wallace, J. (2006). "Identification, model updating, and response prediction of an instrumented 15-story steel-frame building." *Earthquake Spectra,* 22(3), 781-802.

Smith, I. F. C., and Kripakaran, P. (2008). "Model based reasoning for life-cycle structural engineering." *IALCCE08,* Varenna, Italy.

Smith, S. W. (1992). "Iterative use of direct matrix updates: Connectivity and convergence." *33rd AIAA Structures, Structural Dynamics and Materials Conference*, Dallas, TX.

Smith, S. W., and Beattie, C. A. (1991). *Model correlation and damage location for large space truss structures: Secant method development and evaluation*, CR-188102, NASA, Washington, DC.

Smyth, A. W., and Betti, R., Eds. (2004). *Proceedings of the IASC fourth international workshop on structural control*, Columbia University, New York, NY.

Smyth, A. W., Pei, J. S., and Masri, S. F. (2003). "System identification of the Vincent Thomas suspension Bridge using earthquake inputs." *Earthquake Engineering and Structural Dynamics*, 32, 339-367.

Spanos, P. D., and Zeldin, B. A. (1998). "Generalized random decrement method for analysis of vibration data." *Transactions of the ASME*, 120, 806-813.

Spitznogle, F. R., and et al. (1971). *Representation and analysis of sonar signals, vol. 1: Improvements in the complex exponential signal analysis computational algorithm*, U1-829401-5, N00014-69-C0315, Office of Naval Research, Arlington, VA.

Spitznogle, F. R., and Quazi, A. H. (1970). "Representation and analysis of time-limited signals using a complex exponential algorithm." *Journal of the Acoustical Society of America*, 47(5), 1150-1155.

Tamura, Y., and Suganuma, S. (1996). "Evaluation of amplitude-dependent damping and natural frequency of buildings during strong winds." *Journal of Wind Engineering and Industrial Aerodynamics*, 59(2-3), 115-130.

Tan, R. Y., and Cheng, W. M. (1993). "System identification of a non-classically damped linear system." *Computers and Structures*, 46, 67-75.

Turek, M., and Ventura, C. E. (2007). "A method for implementation of damage detection algorithms for civil SHM systems." *3rd International Conference on Structural Health Monitoring and Intelligent Infrastructure*, Vancouver.

Van Overchee, P., and de Moore, B. (1996). *Subspace identification for linear systems: Theory, implementation and applications*, Kluwer Academic Publishers, Dordrecht, Netherlands.

Vandiver, J. K., Dunwoody, A. B., Campbell, R. B., and Cook, M. F. (1982). "A mathematical basis for the random decrement vibration signature analysis technique." *Journal of Mechanical Design*, 104, 307-313.

Vold, H., Kundrat, J., Rocklin, G. T., and Russel, R. (1982). *A multi-input modal estimation algorithm for mini-computers*, 820194, SAE International, Warrendale, PA.

Vold, H., and Rocklin, G. T. (1982). "The numerical implementation of a multi-input modal estimation method for mini-computers." *1st International Modal Analysis Conference*, Society of Experimental Mechanics, Orlando, FL.

Wang, X., Hu, N., Fukunaga, H., and Yao, Z. H. (2001). "Structural damage identification using static test data and changes in frequencies." *Engineering Structures*, 23(6), 610-621.

Werner, S. D., Beck, J. L., and Levine, B. (1987). "Seismic response evaluation of Meloland Road overpass using 1979 Imperial Valley earthquake records." *Earthquake Engineering and Structural Dynamics*, 15, 249-274.

Worden, K., Farrar, C. R., Manson, G., and Park, G. (2007). "The fundamental axioms of structural health monitoring." *Philosophical Transactions of the Royal Society of London, Series A, Mathematical and Physical Sciences*, 463(2082), 1639-1664.

Worden, K., and Lane, A. J. (2001). "Damage identification using support vector machines." *Smart Materials and Structures*, 10(540), 547.

Wu, M., and Smyth, A. W. (2007). "Application of the unscented Kalman filter for nonlinear structure system identification." *Structural Control and Health Monitoring*, 14, 971-990.

Yao, J. T. P., and Natke, H. G. (1994). "Damage detection and reliability evaluation of existing structures." *Structural Safety*, 15, 3-16.

Yoon, S. W., and Ju, Y. K. (2004). "Dynamic properties of tall buildings in Korea." *CTBUH 2004*, Seoul, Korea.

Yoshida, I. (2001). "Damage detection using Monte Carlo filter based on non-Gaussian noise."Newport Beach, CA.

Yoshida, A., and Tamura, Y. (2005). "Response monitoring of tall structures and modal identification." *6th Asia-Pacific Conference on Wind Engineering (APCWE-VI)*, Seoul, Korea.

Yoshida, A., Tamura, Y., Tsuruga, T., and Itoh, T. (2004). "System identification of structure for wind-induced responses." *International Conference on Noise and Vibration Engineering (ISMA 2004)*, Leuven, Belgium.

Yun, C. B., and Shinozuka, M. (1980). "Identification of nonlinear structural dynamic systems." *Journal of Structural Mechanics*, 8(2), 187-203.

Zhang, H., Foliente, G. C., Yang, Y., and Ma, F. (2002). "Parameter identification of inelastic structures under dynamic loads." *Earthquake Engineering and Structural Dynamics*, 31(5), 1113-1130.

Zhang, L., Kanda, H., Brown, D. L., and Allemang, R. J. (1985). "A polyreference frequency domain method for modal parameter identification." *ASME*, 85-DET-106, 6.

Zimmerman, D. C., and Kaouk, M. (1992). "Eigenstructure assignment approach for structural damage detection." *AIAA Journal*, 30(7), 1848-1855.

Zimmerman, D. C., and Kaouk, M. (1994). "Structural damage detection using a minimum rank update theory." *Journal of Vibration and Acoustics*, 116, 222-230.

Zimmerman, D. C., and Smith, S. W., Eds. (1992). *Model refinement and damage location for intelligent structures*, Kluwer Academic Publishers, Dordrecht, Netherlands.

Chapter 6
Utilization of St-Id for Assessment and Decision Making

6.1 Introduction

Decision-making, especially related to structural maintenance, preservation or replacement occurs at a complicated intersection between technical aspects and social, political, environmental, and economic considerations. Fully appreciating the complex context in which the results of a St-Id will be used is critical to developing and executing successful applications. For publicly owned structures, decisions are necessarily made through or influenced by politics. For privately owned structures, such as buildings, the formal political process is less relevant, but non-technical considerations remain substantial. For example, many building owners are concerned about how interventions or investigations into structural performance are viewed by the general public, as impressions of safety or serviceability concerns may negatively influence a building's market or rental values.

Although aiding in the decision-making process is the goal of many St-Id applications, the reality is that sound decision-making requires many non-technical influences to be incorporated and balanced. For the vast majority of infrastructure decisions, technical information may have limited influence. On the other hand, technical information, especially if related to safety concerns, does have the unique ability to trump any other issue and almost exclusively drive decisions. It follows that the influence and value of technical information within the decision-making process is directly related to what the information indicates. If a St-Id reveals expected performance, its influence may be limited; if it indicates high vulnerability to failure, it may overwhelm all other considerations.

In addition, regardless of what the technical information indicates, it is critical that it be presented in a clear manner and that it is relevant to the issue being deliberated. While this may appear obvious at first glance, it is the writers' experience that this issue is not always fully appreciated in practice. For example, consider the case where a bridge is posted for lower than legal loads and a St-Id is carried out to estimate its load carrying capacity. If one were to follow procedures for a diagnostic-level load test, critical members would be identified (using an a priori model) and their responses under known truck loads would be recorded. This information would then be used to 're-rate' these members; in many cases reducing the demand relative to that predicted by the a priori model. Since the rating factors of the critical members are improved, the overall rating of the structure increases.

While this use of experimental data may appear reasonable, the data is not being used properly and thus the results cannot reliably inform decisions regarding the capacity of the bridge. By examining this procedure critically for cases where the load test results can justify an increase in the rating factor, it becomes apparent that the a proper interpretation of the results indicates two things: (1) the a priori model is not representative of the structure (since it significantly disagrees with the

114   STRUCTURAL IDENTIFICATION OF CONSTRUCTED SYSTEMS

experimental results), and thus should be correlated with experimental results through a St-Id before any of its output is trusted, and (2) the members and action (i.e., moment) that were thought to cause overstressing, do not. While these outcomes are useful, they do not provide insight into the actual capacity of the bridge. To do this the model must be reconciled with comprehensive response data capable of providing information about key sources of uncertainty within the model, not just the responses of a few members deemed critical by an erroneous a priori model.

In summary, the influence and value of technical information (obtained through a St-Id) within the decision-making process is related to what it indicates, how relevant it is to the issue being deliberated and how clearly it is presented. Given these requirements, it is clear that for a St-Id to influence the decision-making process, each step is critical. By the time the last step is reached the value of the St-Id for decision-making has already been established.

6.2   Performance-Based Engineering

Over the last decade there has been a significant amount of work in the area of performance-based engineering for constructed systems. While a lot of this attention has been focused on design, this work also provides the proper framework and context for decision-making related to existing constructed systems. By defining the key limit states and desired performances of constructed systems from a holistic standpoint, as opposed to prescribing detailed requirements and relying on a process-based approach to design and renewal, performance-based engineering has the potential to revolutionize all aspects of civil engineering practice. Table 6-1 is taken from a report of the ASCE Performance-Based Engineering Committee and illustrates the key goals for constructed systems for various limit states and various perspectives (domains).

Table 6-1 Multi-dimensional performance-matrix for constructed systems

Domain / Limit state	Engineered	Human-societal	Natural
Operational and utility	Safety Security Efficiency	Transparency Organizational effectiveness Fiscal prudence	Sustainability Minimal impact Hazards risks Management
Engineering	Serviceability Durability Safety Stability of failure	Inspectable Maintainable Adaptable Renewable	Recyclable Carbon footprint Unobtrusive
Societal goals	Long-term economic sustainability Preserve culture	Healthy and just society Promote good governance	Respecting the environment Rely on sound science

Source: Aktan et al. (2007); reproduced with permission.

STRUCTURAL IDENTIFICATION OF CONSTRUCTED SYSTEMS   115

First, it is interesting to note how few of these goals are traditionally considered to be technical in nature. As discussed previously, it is critical to balance both technical and non-technical concerns/perspectives within the decision-making process, and the paradigm of performance-based engineering naturally does this. Second, consider how many of these goals can actually be informed or achieved through applications of St-Id. The goals shown in Table 6-1 are representative of common concerns of owners that have been influenced either directly or indirectly by applications of St-Id. Examples of direct influences include structural safety, stability of failure, hazards risk management, etc., where applications of St-Id have the ability to provide information directly related to the relevant goals. In the case of indirect influences, St-Id applications have the ability to contribute to meeting goals, but in an indirect way. Examples of this type of influence include organizational effectiveness, transparency, and maintainability, among others.

6.3 Risk-Based Decision Making

While the paradigm of performance-based engineering provides the holistic viewpoint necessary for effective decision-making, a versatile metric which can be used to measure the degree of performance (or lack of performance) related to the goals shown in Table 6-1 is also needed. Perhaps the most common metric in this regard is risk, as it is both versatile and easily understood by technical and non-technical audiences. The use of risk to inform decisions dates back many decades and has proven useful in a diverse range of fields such as finance, medicine, insurance, etc. The typical definition of risk is the probability of an event occurring (such as a structural failure) times the consequences associated with the event. By addressing both the probability and consequences of an event, risk is a metric that can balance a diverse set of goals and on which many decision-makers intuitively identify with.

Regardless of whether they are aware of the vast literature related to risk, it is the writers' experience that most structure owners inherently take a risk-based approach to decision-making. As such, it is important that any St-Id application be conceived and carried out from a risk assessment standpoint. In general terms, owners are concerned with the following three questions:

1. What is the probability that the structure will fail to perform, $p(f)$, related to operations/maintenance, serviceability/durability, or safety; (hazard)(vulnerability)?

2. What are the consequences associated with the structure failing to perform; (exposure)?

3. How certain are the estimates to questions (1) and (2); (uncertainty premium)?

For the following discussion, the common definition of risk (the product of the probability of failure [Question 1] and the consequences of failure [Question 2]) will be referred to as the actual risk. In practice, there is an additional component termed uncertainty premium, which is related to Question 3 above and represents the

116 STRUCTURAL IDENTIFICATION OF CONSTRUCTED SYSTEMS

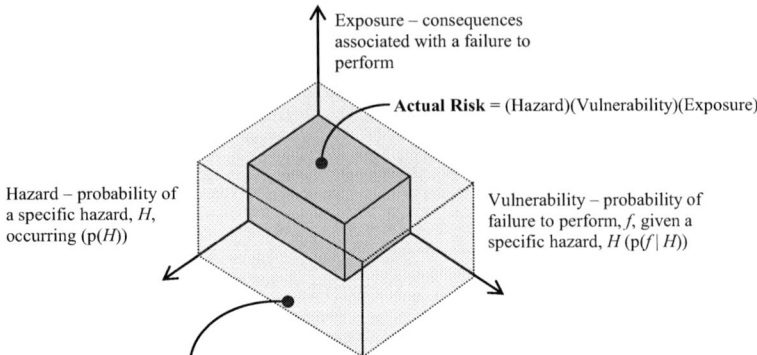

Figure 6-1 Schematic representation of actual and preceived risk

uncertainty associated with the estimates of the other components of risk. This uncertainty premium is what distinguishes the actual risk from the perceived risk.

To visually illustrate both perceived and actual risk, and their individual components, consider Figure 6-1. This figure shows three axes which correspond to the three components of risk typically used in civil/structural engineering: hazard, vulnerability, and exposure. The volume of the box defined by these components represents the risk posed by a structure for the various hazards considered. Based on the writers' experience, the perceived risk (inclusive of the uncertainty premium) is generally much larger than the actual risk, owing to the conservative nature of infrastructure decision-makers. There are notable exceptions to this rule; most obviously these include bridge failures in which owners would have never, knowingly, assumed the actual risk.

Given this context, the role of St-Id in decision-making can be seen as the reduction of the uncertainty premium. With the focus of St-Id being the better understanding of the performance of constructed systems, this reduction in the uncertainty premium afforded by St-Id is nearly always related to better estimates of vulnerability. As a result, if one were to examine the common objectives set forth by the owners (see Chapter 1), nearly all of them seek a better estimate of vulnerability. Notable exceptions to this rule may be cases where the principal uncertainty is related to the level of hazards, such as wind pressure on high-rise buildings or truck weights, speed and frequency for bridges. Table 6-2 provides some examples of hazards, vulnerabilities and exposures for various limit states of constructed systems.

STRUCTURAL IDENTIFICATION OF CONSTRUCTED SYSTEMS

Table 6-2 Some relevant performance limit states, hazards, vulnerabilities and exposures with regard to bridges

Performance Limit States	Hazards	Vulnerabilities	Exposures

Safety: Geotechnical/ Hydraulic	• Flowing water
	• Debris and ice
	• Seismic
	• Vessel Collision
	• Flood
		• Scour/Undermining
		• Loss of support
		• Soil liquefaction
		• Unseating of superstructure
		• Settlement
		• Overtopping
			• Loss of human life
			• Replacement and repair costs
			• Impact of removal from service related to:
			  • Safety – life line,
			  • Economic
			  • Social – mobility
			  • Defense

Safety: Structural	• Seismic • Repeated loads • Trucks and overloads • Vehicle collision • Fire	• Lack of ductility and redundancy • Fatigue and fracture • Overloads • Details and bearings	
Serviceability, Durability and Maintenance	• Winter maintenance practices • Climate • Intrinsic Loads • Impact (Vertical) • Environment	• Corrosion • Cracking/spalling • Excessive deflections/ vibrations • Chemical attacks/reactions • Difficulty of maintenance	• User costs • Maintenance costs • Direct • Indirect – delays, congestion, etc.
Functionality and Cost	• Traffic • Special traffic and freight demands	• Network redundancy and adequacy • Geometry and roadway alignment	• Loss of human life and property (accidents) • Economic and social impacts of congestion

Source: Moon et al. (2010); reproduced with permission.

6.4 Quantitative vs. Qualitative Risk Assessment

Although the formal definitions of risk and vulnerability are useful from a conceptual standpoint, it is almost impossible to estimate them in absolute, quantitative terms. To accomplish this, a model which is not only representative of the structure as observed during testing, but also at the limit state of interest is required. If the structure's behavior at the limit state of interest is similar to the observed response of the structure during testing, it may be possible to develop such quantitative estimates. Such a case may involve a serviceability concern and a St-Id that monitored the structure during normal service conditions.

However, for ultimate limit states, a St-Id application may not justify the effort required to develop quantitative estimates of vulnerability and risk. Consider the case where an ambient vibration monitoring is used to inform seismic vulnerability. The experimental data acquired would be representative of the low-amplitude, linear response of the structure. This data will carry very little information related to the higher-amplitude, nonlinear response, which will no doubt be mobilized during the considered earthquake. This is not to suggest that there is no value in such an application, just that the remaining uncertainty is so significant that the additional effort to develop quantitative estimates of vulnerability and risk is not justified. In these cases, providing owners with quantitative vulnerability and risk estimates is also irresponsible as it conveys much greater certainty than is possible.

In the writers' experience, the majority of the St-Id applications fall into this latter category where quantitative risk and vulnerability estimates are inappropriate. The reality is that such cases demand experience, transparency, and insight to properly convey what the St-Id application indicates and what it does not indicate in an honest and useful manner. The concepts of vulnerability and risk remain critical in these situations, but must be framed in a qualitative, relative sense.

6.5 Closing Remarks

While this report defines and describes the six steps of St-Id (described in the first chapter), the reader is cautioned that this is not a process to be followed like a prescriptive code. The civil engineering profession has a propensity for procedures that do not rely on the talent, experiences and creativity of engineers; the paradigm of St-Id does not fall into this category. From the outset and through each of the six steps described, successful applications require diligence and wide-ranging expertise that is typically not offered at universities and is not found in any text book. Achieving this diversity of knowledge requires a multi-disciplinary team led by individuals that have sound heuristic-knowledge of the constructed system of interest. There is a danger that when new technologies or paradigms are introduced into a field, age-old heuristics and wisdom are either forgotten or under-appreciated – in the case of St-Id, this would obviate nearly all of its potential.

6.6 References

Aktan, A. E., Ellingwood, B., and Kehoe, B. (2007). "Performance-based engineering of constructed systems." *Journal of Structural Engineering,* 133(3), 311-323.

Moon, F. L, Frangopol, D. M., Catbas, F. N., Aktan, A. E. (2010) "Infrastructure Decision-Making based on Structural Identification," Proceedings, ASCE Structures Congress, p590-596, Orlando, FL, May 12-14.

Appendix A
Case Studies on the Structural Identification of Buildings

This chapter presents case studies documenting the successful application of system identification to constructed buildings, as discussed in two recent reviews (T. Kijewski-Correa and Cycon 2007; T. Kijewski-Correa et al. 2008). These case studies are intended to validate and improve the design state-of-the-art, subject to the constraints of their monitoring hardware and generally do not execute step 6 of the structural identification process. An inventory of instrumented buildings compiled by these reviews is offered by Table A-1, and since many applications are a result of California Strong Motion Instrumentation Program (CSMIP), this program will receive additional attention in Section 7.3. In fact, strong motion programs have generated more than 150 instrumented buildings in the United States (Huang and Shakal 2001), 100 sites in Japan (Lin et al. 2003) and 40 sites in Taiwan (Huang 2006). This is in stark contrast to the situation outside of seismic zones, especially in the United States, where owners fear that public disclosure of monitoring efforts may generate public misconceptions regarding the building's condition and even liability issues (T. Kijewski-Correa and Kareem 2007).

A.1 Chicago Full-Scale Monitoring Program

A.1.1 Program Description

Long before the notion of *Performance Based Engineering* was popularized in seismic circles it had been practiced in the design of tall buildings under wind, where survivability, serviceability and even habitability limit states must be simultaneously evaluated. The assessment of structural performance under these varying limit states using full-scale monitoring has been commonplace for over a decade, particularly in Japan, for the purpose of validating/operating supplementary energy dissipation devices to limit accelerations (Kareem et al. 1999). Still the aim of these efforts was not long-term evaluations of performance or validation of the underlying design process. Subsequently, Brownjohn et al. (1998) began tracking the response variations of Republic Plaza in Singapore, while a systematic effort to validate the design state-of-the-art for tall buildings was undertaken through the Chicago Full-Scale Monitoring Program (see windycity.ce.nd.edu). The latter project has been comparing the full-scale accelerations and displacements of a collection of tall buildings in Chicago to the wind-tunnel response predictions used in their design since 2002 (T. Kijewski-Correa et al. 2006). The program has since expanded to now include buildings overseas (T. Kijewski-Correa and Pirnia 2007); however, the present case study focuses on the original three buildings in Chicago, whose names were withheld at the request of the owners:

120 STRUCTURAL IDENTIFICATION OF CONSTRUCTED SYSTEMS

Table A-1. Partial inventory of instrumented buildings in the literature

Building	Location	Ht. [m]	No. of Sensors/ Recorded Events
A-Chicago [3 Bldgs]	Chicago	--	4/AV since 2002
Anonymous		7-ST	29/Chi-Chi EQ
Anonymous		13-ST	15/Whittier Narrows EQ (1987)
Anonymous	Viña del Mar, Chile	22-ST	Central Chile EQ (1985)
Anonymous	Hong Kong	120	2/AV since 1995
Anonymous	San Francisco	172	18/Loma Prieta EQ (1989)
Anonymous	Los Angeles	221	20/Northridge EQ (1994)
Anonymous	Boston	245	8/AV 1973-1978
Anonymous	Seoul	264	6/AV since 2005
Alhambra Bldg.	Los Angeles	13-ST	12/Asst. EQs
Atwood Bldg.	Anchorage, AK	80.5	32/Asst. EQs
Bank of California	Los Angeles	12-ST	9/San Fernando EQ (1971)
Bank of China	Hong Kong	370	2/T. Sally (1996)
Building C	Hong Kong	218	2/T. Imbudo (2003), T. Dujuan (2003)
Building E	Hong Kong	206	2/T. Imbudo (2003), T. Dujuan (2003)
Central Plaza Tower	Hong Kong	374	2/T. Sally (1996)
CSUH Admin. Bldg.	Hayward, CA	61	16/Loma Prieta EQ (1989)
Di Wang Tower	Shenzen, PRC	384	2/T. Sally (1996)
FCC Bldg.		9.75	16/Northridge EQ (1994)
Great Western Savings and Loan	San Jose, CA	10-ST	8/Morgan Hill EQ(1984)
Guangdong Intl. Bldg.	Guangzhou, PRC	200	2/AV
Imperial Co. Services Bldg.	El Centro, CA	6-ST	13/Imperial Valley EQ (1979)
Jin Mao Tower	Shanghai	365	2/T. Rananim (2004)
Law & Justice Center	R. Cucamonga, CA	4-ST	19/Redlands EQ (1985)
Millikan Library	Pasadena, CA	43.9	36/Whittier Narrows EQ (1987)
Pacific Park Plaza	Emeryville, CA	94	21/Loma Prieta EQ (1989)
Republic Plaza	Singapore	280	4/21 Minor EQ
Office Bldg.	San Bruno, CA	24	13/Loma Prieta EQ (1989)
Santa Clara Co. Office Bldg	San Jose, CA	57	23/Morgan Hill EQ (1984), Loma Prieta EQ (1989)
Sylmar Co. Hospital		6-ST	13/Northridge EQ (1994)
Transamerica Bldg.	San Francisco	257	22/Loma Prieta EQ (1989)
Town Park Towers	San Jose, CA	10-ST	6/Morgan Hill EQ (1984)
UCLA Bldg.	Los Angeles	66	72/Parkfield EQ (2004)
Union Bank Bldg.	Los Angeles	42-ST	2/San Fernando EQ (1971)
Van Nuys Hotel	Van Nuys, CA	7-ST	16/Big Bear EQ (1992), Northridge EQ (1994)
Database [205 Bldgs.]	Japan	< 300	≥2/EQ, AV, Forced Vibration
Database [67 Bldgs.]	Korea	< 243	≥2/AV
Database [185 Bldgs.]	Worldwide	< 337	≥2/AV

Notes: ST: Stories, AV: Ambient Vibrations, EQ: Earthquake, T: Typhoon
Source: Adapted from Kijewski-Correa and Cycon (2007).

- Building 1: The primary lateral load-resisting system features a steel tube comprised of exterior columns, spandrel ties and additional stiffening elements to achieve a near uniform distribution of load on the columns across the flange face, with very little shear lag. As such, lateral loads are resisted primarily by cantilever action, with the remainder carried by frame action.

- Building 2: In this reinforced concrete building, shear walls located near the core of the building provide lateral load-resistance. At two levels, this core is tied to the perimeter columns via reinforced concrete outrigger walls to control the wind drift and reduce overturning moment in the core shear walls.

- Building 3: The steel moment-connected, framed tubular system of Building 3 behaves fundamentally as a vertical cantilever fixed at the base to resist wind loads. The system is comprised of closely-spaced, wide columns and deep spandrel beams along multiple frame lines. Deformations of the structure are due to a combination of axial shortening, shearing and flexure in the frame members, and beam-column panel zone distortions.

A.1.2 Objectives of St-Id Application

The Chicago Full-Scale Monitoring Program has the overarching goal of providing the first systematic full-scale validation of tall building design practice in this country. Specific objectives of the program include:

- Development and introduction of advanced instrumentation systems, including GPS, for in-situ structural monitoring, real-time data access and analysis via a Java-Based web interface, promoting the use of full-scale monitoring of tall buildings in the United States.

- Establishment of real-time monitoring of tall building response during significant wind events with measured wind speed and direction, providing a comparison of actual response to predicted response estimates

- Investigation of the sensitivity of in-situ dynamic characteristics of tall buildings based on the level of response, foundation type, materials of construction, age and condition of the building, and wind environment

A.1.3 Model Development

For the design of tall buildings under wind, two types of models are required in the design stage, both of which were validated in the Chicago Full-Scale Monitoring Program. These two model types are now described.

Finite Element Model: Designer estimates of viscous damping ratios, natural frequencies, and mode shapes, in conjunction with base-moment information from the wind tunnel tests, are required to estimate accelerations of tall buildings under wind. Thus the development of finite element models is first addressed. For the buildings associated with this study, finite element models were developed at

Skidmore Owings and Merrill LLP using currently available commercial software: ETABS and SAP 2000, based upon careful reference to the design drawings. It was not the purpose of this study to apply a unique set of modeling assumptions to this process in order to mimic a known, in-situ measured result. Nor was model updating desired. Rather, all assumptions regarding the finite element representation of the buildings in this study reflect those commonly applied in design offices for serviceability assessment. This then allows the discrepancies between common designer assumptions and in-situ behavior to be underscored.

The mass associated with the self-weight of the structure and the full weight of the exterior cladding system are included in the dynamic analysis. Additionally, special attention is paid to the use of the building and the resulting loading conditions at each floor in order to determine what fraction of the design imposed load to include in the mass calculations for the dynamic analysis. Due to the heights of the study buildings, the analysis includes the effects of building displacement on the frequencies through a second-order (P-Δ) iteration. The buildings were modeled as fixed at the base, such that no base rotation exists. This is thought to approximate the generally high soil-structure interfacial stiffness observed in Chicago for buildings under transient lateral loads.

For Buildings 1 and 3, framed primarily in structural steel, the representation of the member stiffness was straight-forward, as the steel elements remain elastic at service level loadings. For the reinforced concrete building (Building 2), adjustments were made to selected lateral-load resisting elements to represent the post-cracking stiffness of these elements under service level loads. Specifically, the flexural and shearing stiffnesses of the link (coupling) beams within the shear wall system were reduced to one-half and one-fifth of the elastic stiffness, respectively. The beam-supported slab was modeled using shell elements. The flexural stiffness of the slab's shell elements was set to one-half of the elastic stiffness in order to approximate the post-cracking behavior of the slab, which transfers flexure and shear between the perimeter columns and core shear walls. While generally considered to support gravity floor loads alone, explicit modeling of the linkage between the floors, exterior columns, and core often results in a substantial contribution to lateral resistance in reinforced concrete buildings.

<u>Wind Tunnel Models:</u> The model-scale wind-induced responses of the three buildings selected for the study were measured in the high-speed section of the closed-circuit wind tunnel (BLWT II) at the Boundary Layer Wind Tunnel Laboratory (BLWTL) at the University of Western Ontario (UWO). The high-frequency force-balance (HFFB) method was chosen for the wind tunnel tests as it allows the flexibility to repeat response predictions (accelerations) based on the measured modal force spectra but considering different building dynamic properties without the requirement of additional wind tunnel testing. Accordingly, differences between the in-situ and predicted structural properties of the buildings are easily reconciled using the HFFB method. The modeling for the force balance tests conducted in this study consisted of three components: 1) a rigid and lightweight detailed 1:500 scale model of each of the study buildings; 2) a detailed model of the structures surrounding the building sites within a full-scale radius of about 750 m;

STRUCTURAL IDENTIFICATION OF CONSTRUCTED SYSTEMS 123

and 3) a less detailed model of the upstream terrain, chosen to simulate the scaled turbulence intensity and velocity profiles expected at full-scale for each site. Each building was then tested in this environment at 10° increments for the full 360° azimuth range. Time histories of the responses, as well as the mean and RMS base bending and torsional moments were recorded and their associated power spectra were subsequently obtained. The generalized forces acting on the building in the sway directions are related to the base moments and then used to determine RMS accelerations in the two sway and torsional responses using random vibration analysis (T. Kijewski-Correa et al. 2006).

A.1.4 Experimental Studies

The basic instrumentation array consists of orthogonal pairs of force-balance accelerometers installed in opposite corners of the uppermost floor of each building, as shown in Figure A-1, interrogated by an on-site datalogger that streams the data via DSL modem to an off-site server. Recognizing the importance of mean and background (quasi-static) response components, Building 1 was supplemented with a high precision global positioning system. In addition, due to the variability of the wind field in urban zones, wind velocities are recorded by ultrasonic anemometers on the rooftop of Buildings 1 and 3, as well as at the surrounding airports and on the lake front. Responses of the buildings are recorded under naturally occurring vibrations, primarily wind-induced, with 10-minute statistics continuously archived and one hour time histories sampled at 10 Hz whenever responses surpass a particular threshold. The primary instrumentation systems were respectively installed in Buildings 1, 2 and 3 on 06/14/02, 6/15/02 and 4/30/03.

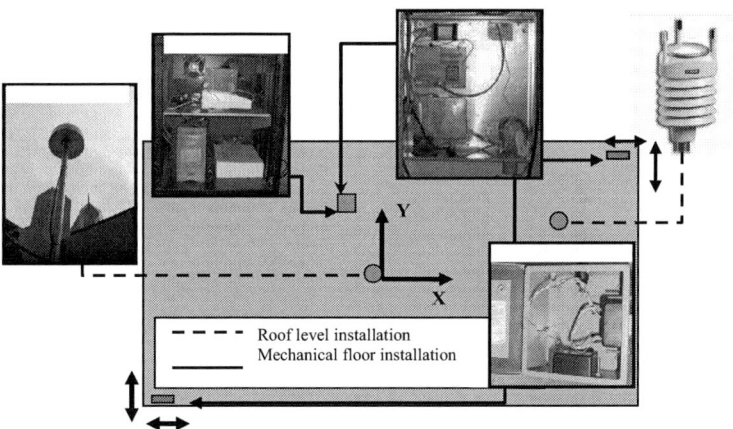

Figure A-1 Schematic representation of sensors used in the Chicago Full-Scale Monitoring Program (Source: Kijewski-Correa et al. 2006; reproduced with permission)

A.1.5 Data Analysis and Model Calibration

Dynamic Properties: The validation sought in this case study is two-fold. First, since habitability accelerations are the primary concern in tall buildings and are particularly sensitive to dynamic properties specified by designers, modal properties estimated by Random Decrement Technique (RDT) with local averaging were used to verify the validity of common assumptions used by designers (T. Kijewski-Correa et al. 2006). This form of local averaging has been shown to improve RDT performance by minimizing trigger sensitivity, while still permitting documentation of the degree of amplitude dependence in viscous damping ratios of several tall buildings in Chicago, as well as an accompanying CoV to document reliability (T. Kijewski-Correa and Pirnia 2007). A comparison between the in-situ and assumed/predicted dynamic properties is provided in Table A-2 for the two fundamental lateral modes. Due to the relatively high torsional stiffness of the buildings, response levels in this direction were relatively modest and often scarcely above the noise floor. Although cantilever-dominated steel structures (Building 1) were found to have near perfect agreement with design-predicted fundamental lateral frequencies, reinforced concrete interactive systems (Building 2) were found to be stiffer in-situ (by up to 25%), which may be attributed to differences in in-situ modulus of elasticity or model stiffness reductions due to cracking that has yet to be observed in the service life of the building. Conversely, interactive steel systems (Building 3) deviated from FEM predictions, yielding longer in-situ periods by approximately 10%, potentially due to unmodeled panel zone deformations and foundation interactions, as explored in separate sensitivity studies (Bentz et al. 2010; T. Kijewski-Correa et al. 2005).

Table A-2. Comparison of predicted/assumed dynamic properties and those observed in full-scale for three buildings in the Chicago Full-Scale Monitoring Program

		Building 1	Building 2	Building 3
Design Period [s]	X-Sway	4.9	6.7	7.7
	Y-Sway	7.0	6.5	7.6
In-Situ Period [s]	X-Sway (COV)	4.89 (0.10%)	5.61 (0.22%)	8.60 (0.25%)
	Y-Sway (COV)	7.11 (0.19%)	5.66 (0.68%)	8.60 (0.14%)
Design Damping	X-Sway	1%	1%	1%
	Y-Sway	1%*	1%*	1%*
In-Situ Damping	X-Sway (COV)	0.87% (23.9%)	1.42% (7.4%)	1.04% (20.6%)
	Y-Sway (COV)	0.88% (8.9%)	2.4% (8.0%)	1.21% (23.0%)

*1% used for accelerations, 1.5% used for base moments.
(Source: Kijewski-Correa et al. 2006; reproduced with permission)

STRUCTURAL IDENTIFICATION OF CONSTRUCTED SYSTEMS 125

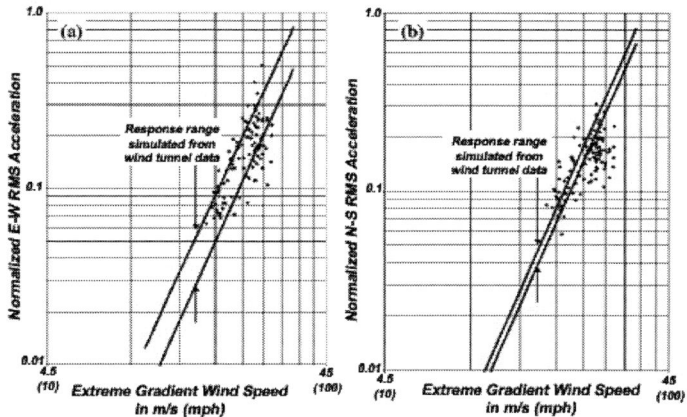

Figure A-2 Comparison of wind tunnel predictions and in-situ RMS accelerations for Building 2 of the Chicago Full-Scale Monitoring Program (Source: Kijewski-Correa et al. 2006; reproduced with permission)

Wind-Induced Accelerations: In order to isolate the effect of uncertainty in dynamic properties, comparisons between the full-scale accelerations and predictions based on force balance wind tunnel tests were conducted, using the in-situ dynamic properties observed. Response predictions were made for a range of wind angles and in particular, damping levels, based on the variation observed in the estimates in Table A-2. Due to the spatio-temporal variability of wind, comparisons of RMS accelerations utilized wind tunnel predictions for developed boundary layer flows with comparable turbulence intensity, the same mean wind speed and range of angles observed in a given wind event. While generally good agreement was observed, under and overestimates were observed for the suite of buildings over a variety of wind events of varying intensity and angle of attack. An example of this type of comparison is shown in Figure A-2 for a spring event affecting Building 2. The scatter observed can in part be attributed that the recorded wind data is not necessarily representative of conditions at each building. Similar comparisons were also made for the displacements of Building 1 using the GPS sensor (T. Kijewski-Correa and Kochly 2007), noting similar trends for the resonant response component.

A.1.6 Interpretation

This and other full-scale investigations have also been used to help formulate hypotheses surrounding energy dissipation capabilities tied to dominant deformation mechanisms in these systems (Bentz and Kijewski-Correa 2008). Based on data collected in this study, it has been reaffirmed that concrete in general dissipates more energy than steel, but more importantly that structural systems with increasingly more cantilever action manifest less viscous damping (as shown in Figure A-3) and show less amplitude dependence. This can be seen from the comparison of damping levels

126   STRUCTURAL IDENTIFICATION OF CONSTRUCTED SYSTEMS

in Building 1 and Building 3, both made of steel, and between the two axes of Building 2 where different structural systems are employed, the outrigger system in particular being characterized by more cantilever action.

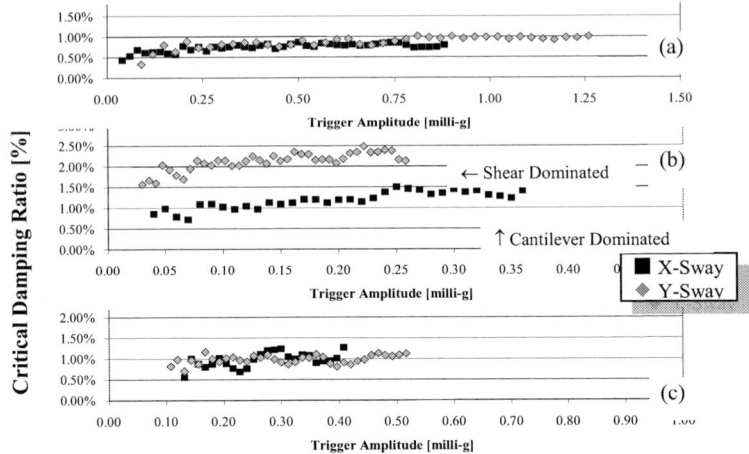

Figure A-3 Amplitude-dependent damping from three tall buildings in Chicago (Source: Kijewski-Correa et al. 2006; reproduced with permission)

A.2   Four Seasons Building

A.2.1   Building Description

The Four Seasons Building, located in Sherman Oaks, California, is a four-story reinforced concrete perimeter special moment frame and an interior post-tensioned slab-column frame with square columns and drop panels constructed in 1977 that was damaged during the 1994 Northridge earthquake (Yu et al. 2007; Yu, Wallace et al. 2007). Bell caissons connected by grade beams constitute the building's foundation. Visual inspection was performed to document the earthquake damage, which included slab punching failures (Figure A-4a), significant diagonal cracks in beam-column joint regions (Figure A-4b), column flexural cracks, and concrete spalling at beam ends adjacent to the beam-column joints. Slated for demolition, it remains unoccupied since that incident. The UCLA/NEES project focused on this building involves forced vibration testing of the building using the NEES equipment as well as detailed system identification and numerical modeling studies.

A.2.2   Objective of St-Id Application

Given that there is a lack of adequately high spatial resolution data sets from field testing of full-scale structures under dynamic loading condition, earthquake engineers have had a limited ability to improve their understanding of the seismic performance

STRUCTURAL IDENTIFICATION OF CONSTRUCTED SYSTEMS   127

of structural and geotechnical systems. This case study responds to that need by providing the opportunity to deploy sensors with high density and impart controlled excitations to provide valuable insight into the performance of this common type of structural system and reveal the reasons for its poor performance during the Northridge earthquake.

 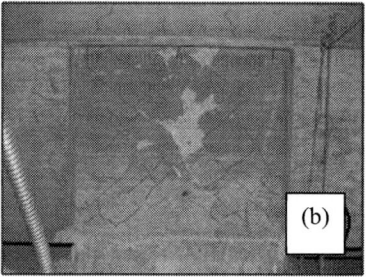

Figure A-4. Observed damage to Four Seasons Building due to (a) slab punching and (b) joint shear cracking (Source: Yu et al, 2007; reproduced with permission from John Wiley and Sons)

A.2.3 Model Development

An initial finite element model was constructed based on architectural/structural drawings and reasonable assumptions. Even though the initial model was constructed with considerable care (Yu et al. 2007; Yu, Wallace et al. 2007), there were significant discrepancies between these identified and analytical modal properties as shown later in Table A-3.

A.2.4 Experimental Studies

A dense instrumentation array was utilized that included 36 accelerometers, 20 displacement transducers, and 94 strain gauges. A wireless local area network was installed to acquire and record waveform data. Figure A-5 shows the typical floor plan and sensor layout. Both earthquake-type and harmonic force histories have been applied to the building at the roof level, and waveforms were recorded to enable evaluations of the complete structural system response, structural component behavior (e.g., slab-column connections and pile caps), and the response of non-structural components (e.g., partitions, piping systems) as well as the influence of these components on the system response. These NEES@UCLA forced vibration tests (NEES@UCLA 2010) were conducted in 2004, using both linear and eccentric mass shakers (labeled as LMS and EMS on Figure A-5, respectively).

128 STRUCTURAL IDENTIFICATION OF CONSTRUCTED SYSTEMS

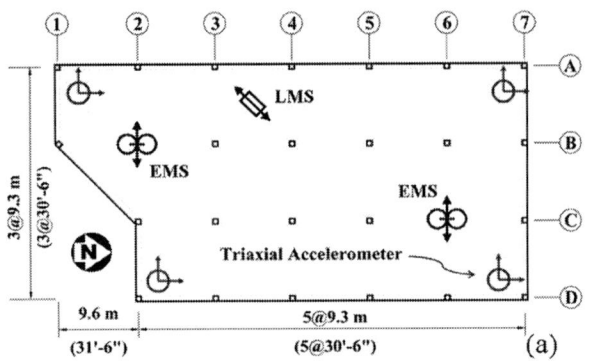

Figure A-5 Typical floor plan of Four Seasons Building showing sensor and shaker locations. (Source: Yu et al, 2007; reproduced with permission from John Wiley and Sons).

A.2.5 Data Analysis, Model Calibration

The Numerical algorithm for Subspace State-Space System Identification (N4SID) was adopted in the analysis of this full-scale dataset (Van Overschee and DeMoor 1994). This non-iterative approach yields reliable state-space models for complex multivariate dynamical systems directly from measured data with modest computational effort. The modal properties are easily deduced from the back-calculated state-space model and are displayed in Table A-3.

Table A-3. Four Seasons Building modal properties

Mode		Identified		Initial Model			Updated Model		
No	Dir	f_m (Hz)	ζ (%)	f_i (Hz)	f_i/f_m	MAC*	f_u (Hz)	f_u/f_m	MAC*
1	EW	0.88	5.66	0.89	1.01	0.98	0.89	1.01	1.00
2	NS	0.94	6.94	1.08	1.15	0.99	0.96	1.02	0.99
3	Tor	1.26	6.01	1.29	1.02	1.00	1.26	1.00	1.00
4	EW	2.73	5.61	2.64	0.97	0.90	2.72	1.00	0.99
5	NS	2.94	7.69	2.99	1.02	0.94	2.93	1.00	0.98
6	Tor	3.44	6.14	3.42	0.99	0.93	3.44	1.00	0.99
*Modal Assurance Criterion (MAC) value denotes a measure of resemblance between two vectors that represent the mode shapes. MAC varies between 0 and 1, with 1 denoting a perfect match. Source: Yu et al. 2007; reproduced with permission from John Wiley and Sons.									

STRUCTURAL IDENTIFICATION OF CONSTRUCTED SYSTEMS 129

Model updating was performed to reduce these discrepancies. As it is well known, this second nonlinear inverse problem is inherently ill-conditioned with non-unique solutions, because groups of distinct model parameters may have very similar influences on the discrepancy residuals. As such, the updating parameters — which included translational masses, gyration radii, and effective beam, column, slab stiffnesses — had to be grouped to reduce the number of independent unknowns. It was not entirely possible to determine a priori what the (pareto-) optimal grouping of parameters would be, so a novel strategy to adaptively constrain the updating parameters was devised and first verified on model problems (Skolnik et al. 2006). This enhanced model updating method was effective, convergent, and yielded reasonable results with improved agreement between identified and computed (from updated FEM) modal data, as further shown in Table A-3. Figure A-6a displays the good agreement between the updated and measured transfer functions for the Four Seasons Building. Figure A-6b shows the initial grouping of the updating parameters that was based on member types and their vertical locations. The aforementioned adaptive constraining algorithm further clustered the updating parameters based on the error function's sensitivity to these groups to yield a convergent solution. A subset of the final values of the updated effective stiffness factors is displayed in Figure A-6c.

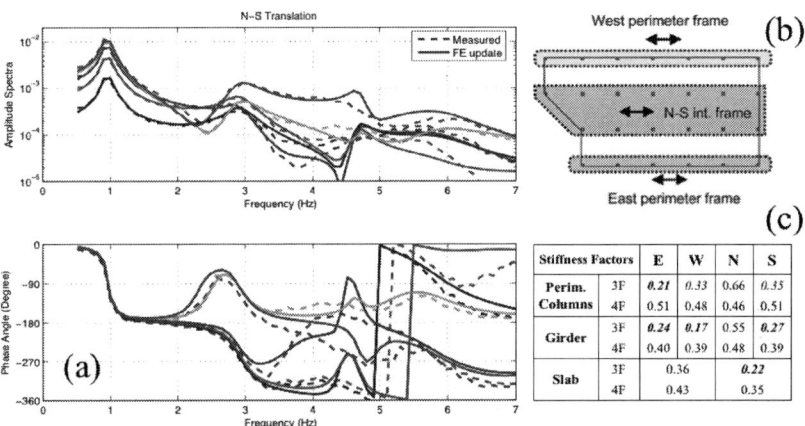

Figure A-6. Four Seasons Building: (a) comparison of measured and predicted transfer functions, (b) sample diagram of initial grouping of updated stiffness parameters, and (c) table of select final effective stiffness factors based on updated parameters (Source: Yu et al, 2007; reproduced with permission from John Wiley and Sons).

130   STRUCTURAL IDENTIFICATION OF CONSTRUCTED SYSTEMS

A.2.6   Interpretation and Decision Making

Several of the updated factors in Figure A-6c were significantly reduced from their initial values of 0.5 (indicated with bold-italic numbers) and corresponded well to locations of observed damage in the building.

A.3   Three-Story Concrete Building in CSMIP

A.3.1   Program Description

California Legislation initiated the California Strong Motion Instrumentation Program (CSMIP) in 1972, in order to establish a sensor network (Figure A-7) for recording seismic events throughout the state of California (California Geological Survey 2007). The stations have been chosen based on the geological characteristics of the site, and the type of the structures in order to create a broad-ranging inventory of seismic records from residential, commercial, healthcare, and industrial facilities. A list of ground response stations with their latitudes and longitudes is available on the strong motion instrumentation program (SMIP) website under the section "Station Information" (California Geological Survey 2007).

The specific case study herein will analyze the strong-motion responses of a three-story concrete building (Figure A-8) and the results are provided as an example that illustrates the potential influence of soil-structure interaction on the dynamic characteristics of a building.

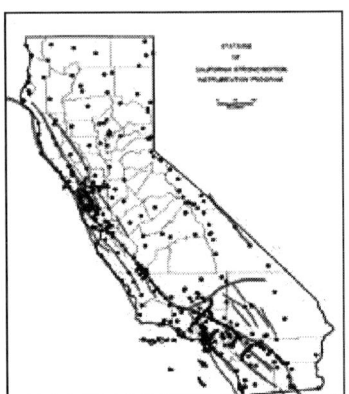

Figure A-7 Distribution of California Strong Motion Instrumentation Program stations, CSMIP (Source: CSMIP)

Figure A-8 Three-story Peidmont School Office Building (Source: Center for Engineering Strong Motion Data operated by CSMIP)

The vertical load carrying system of the specimen building consists of a reinforced concrete frame supporting a concrete pan joist system. Reinforced concrete shear walls resist lateral forces in both transverse and longitudinal directions. The structure is supported by spread footings that are interconnected by reinforced concrete tie beams. Two free-field stations are located less than 800 ft away from the building.

A.3.2 Objective of St-Id Application

One of the main objectives of establishing CSMIP has been to improve the perception of ground motion and seismic response of constructed facilities among researchers in the following ways (California Geological Survey 2007):

- Improvement of the understanding and prediction of the behavior of specific types of buildings during large earthquakes by observing and analyzing the destructive effects of past earthquakes on similar types of buildings;
- Revision of UBC requirements regarding the design of buildings in near-fault zones;
- Assessment of the state-of-health of buildings after an earthquake;
- Improvement of UBC formulations for calculating the fundamental periods of buildings;
- Evaluating the performance of base-isolated buildings during ground shaking.

Another objective is to share related research results with professionals involved in seismic design as well as agencies responsible for post-earthquake planning. In this respect, and through the Data Interpretation Project, which was established in 1989, CSMIP funds projects related to strong-motion analysis and holds annual seminars on the "Utilization of Strong Motion Data." The annual seminar papers, along with reports on earthquake data processing and utilization are available at the SMIP website.

A.3.3 System Identification Method

A well-established method (Safak 1991) suitable for extraction of modal properties is utilized herein based on the parameterization of the structure's discrete-time transfer function defined by

$$H_{ij}(z) = \frac{b_1 z^{-1} + b_2 z^{-2} + \cdots + b_{2n} z^{-2n}}{1 + a_1 z^{-1} + a_2 z^{-2} + \cdots + a_{2n} z^{-2n}} \qquad (4)$$

Equation (4), which represents a discrete-time filter, leads to an ARX model whose coefficients can be solved with a pair of input and output records in hand. For that purpose, the cumulative error between the measured output and the modeled output is minimized in a least-squares sense (Safak 1991). Natural frequencies and modal damping ratios are then extracted from the poles of the discrete-time transfer function (Safak 1991).

132 STRUCTURAL IDENTIFICATION OF CONSTRUCTED SYSTEMS

Treatment of Soil-Structure Interaction: In the presence of soil-structure interaction effects the outcomes of system identification studies are affected by the choice of the input signal. Taking into account the relative motions of the foundation with respect to the ground, Stewart et al. (1998) categorize modal properties of a structure as follows:

- Fixed base properties represent the flexibility characteristics of the structure;
- Flexible base properties represent the flexibility of the structure and the surrounding soil media. The effects of lateral and rocking motions of the foundation are reflected in the flexible base properties;
- Pseudo-flexible base properties are intermediate properties, which represent flexibility characteristics of the structure as well as the effects of base rocking.

The pairs of input-output motions required for identifying the modal properties corresponding to each case are specified in Table A-4. In this table, u_g denotes the ground motion, whereas u_f and u correspond to the translation of foundation and the roof relative to the ground, respectively. The term θ denotes the base rotation; and its product with the effective height (h) of the building yields the roof translation due to base rocking. In order to measure base-rocking in both lateral and transverse directions, it is necessary to instrument the foundation with at least 2 vertical sensors in each direction. Table suggests that the system identification results are not affected by base rocking or translational motion relative to the ground, i.e., soil-structure interaction, if the corresponding motion is included in the input (Stewart and Fenves 1998). For example when both translational and rocking motions of the foundation are included in the input signal, the additional flexibility due to soil-structure interaction is excluded from the results.

Table A-4. Input-output pairs for identifying modal properties associated with varying base fixity

System	Input	Output
Flexible Base	u_g	$u_g + u_f + h\,\theta + u$
Pseudo-Flexible Base	$u_g + u_f$	$u_g + u_f + h\,\theta + u$
Fixed Base	$u_g + u_f + h\,\theta$	$u_g + u_f + h\,\theta + u$

(Source: Stewart and Fenves, 1998; reproduced by permission from John Wiley and Sons)

Due to lack of adequate instrumentation, many of the currently instrumented structures are missing the free-field (u_g) and/or base-rocking (θ) measurements. However, using methods developed by Stewart et al. (1998), it is still possible to estimate one set of modal properties from the other two sets when either the free-field *or* the base-rocking measurements are missing. Table A-5 displays the different cases of missing records and the set of modal properties that may be identified and estimated for each case.

Table A-5. Input-output pairs for identifying modal properties associated with varying base fixity.

Missing Record	Identified Modal Properties	Estimated Modal Properties
u_g	Pseudo-Flexible Base, Fixed Base	Flexible Base
θ	Flexible Base, Pseudo-Flexible Base	Fixed Base

(Source: Stewart and Fenves, 1998; reproduced by permission from John Wiley and Sons)

A.3.4 Instrumentation Overview

The instrumentation utilized at each site in the CSMIP is listed at the program website under the section "Station Information" (California Geological Survey 2007). Examples of some of the instrumentation enclosures are shown in Figure A-9. Specific to this case study, the responses of the three story concrete building were analyzed in three earthquake events recorded by both the building and the free-field sensors. Table A-6 shows the peak accelerations corresponding to each event. In addition to the sensors measuring lateral motions in both the transverse and the longitudinal directions, the structure has been instrumented with two vertical sensors along the north-south (transverse) direction on the ground floor (Figure A-10). Together with the free-field measurements, the building's sensors enable the direct identification of natural frequencies that correspond to all of the various conditions of base fixity in the north-south direction, noting that only date from Channels 2, 3, and 8-11 were used.

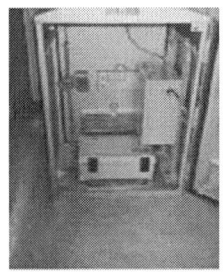

Figure A-9 Typical view of CSMIP instrumentation (Source: California Strong Motion Instrumentation Program)

Table A-6. Recorded peak accelerations of three-story concrete building during three earthquakes

	Peak Acceleration (g)							
	Free Field			Ground Floor			Roof	
Earthquake	NS	EW	V	NS	EW	V	NS	EW
1989 Loma Prieta	0.071	0.084	0.026	0.086	0.072	0.033	0.18	0.15
2007 Lafayette	0.026	0.019	0.018	0.0109	0.006	0.006	0.0245	0.027
2007 Piedmont	0.15	0.12	0.11	0.0678	0.092	0.056	0.144	0.232

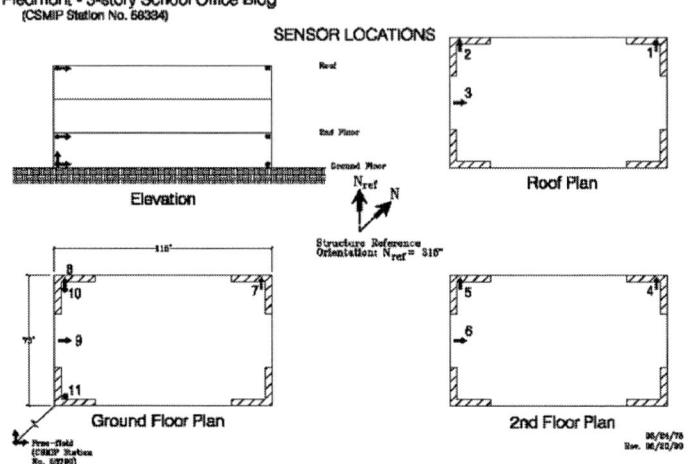

Figure A-10 Sensor layout in three-story concrete building (Source: Center for Engineering Strong Motion Data operated by CSMIP)

A.3.5 Data Processing and Archiving

The primary charge of CSMIP is the collection, processing and archiving of data collected during seismic events. Upon the activation of a station due to ground motions, the CSMIP headquarters is automatically notified; and subsequently, earthquake records are recovered either remotely by modem, or physically at the station. Recorded data are baseline- and instrument-corrected and filtered prior to distribution to the users. Methods employed for digitizing and processing the accelerograms at CSMIP are outlined by Shakal et al. (2004).

Users are usually provided with three sets of numerical data files. Shakal and Huang (1985) provide information regarding the format of strong motion data files. Volume 1 data file contains raw acceleration data (baseline-corrected acceleration records). Volume 2 data file contains processed acceleration, velocity, and displacement records. Volume 3 data file contains Fourier spectrum values, relative displacement, relative velocity, absolute acceleration, and pseudo-velocity response spectrum values for different damping values, as well as time instants of maximum spectral responses (Shakal and Huang 1985). In addition to numerical data, users have access to time history and spectral graphs.

Data can be obtained through Internet Quick Report (IQR) for earthquakes of magnitude 4.0 and above, or through Internet Data Report (IDR) for major earthquakes, or by various searching criteria. All of these three options are available through the website of The US National Center for Engineering Strong Motion Data (CESMD) (CESMD 2011) (Figure A-11a). It is possible to search for data based on the name, magnitude, and date of an event, as well as the location and properties of the structure, peak ground acceleration, and distance from the epicenter (Figure A-11b). Aside from the search tool and Internet reports, data can be obtained directly via the "recorded earthquakes map" or the "station maps" of northern and southern California as illustrated in Figure A-12.

Internet Quick Reports are often supplemented by near-real-time ShakeMaps (Figure A-13), which can be accessed through SMIP website within a few minutes after earthquakes of magnitude 3.5 and higher. Depending on the distance from the epicenter, and non-uniform seismic wave propagation due to geologic factors, various regions experience different levels of shaking during an earthquake. The ShakeMaps present an overall picture of the shaking intensities in the affected areas and provide valuable information for post-event emergency response as well as research purposes. Currently, California Integrated Seismic Network (CISN), which has been funded through California Office of Emergency Services since 2001, is responsible for producing ShakeMaps almost immediately after an earthquake. CISN consists of TriNet, which is a joint project between CSMIP, Caltech, and USGS at Pasadena in southern California. In northern California, CSMIP is collaborating with UC Berkeley and USGS at Menlo Park to contribute to CISN.

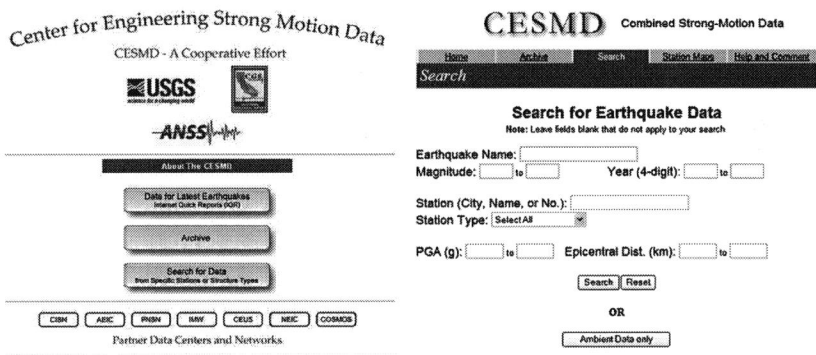

Figure A-11 (left) CESMD IQR, IDR, and search tool; (right) CESMD strong motion search tool. (Source: The Center for Engineering Strong Motion Data operated by CSMIP)

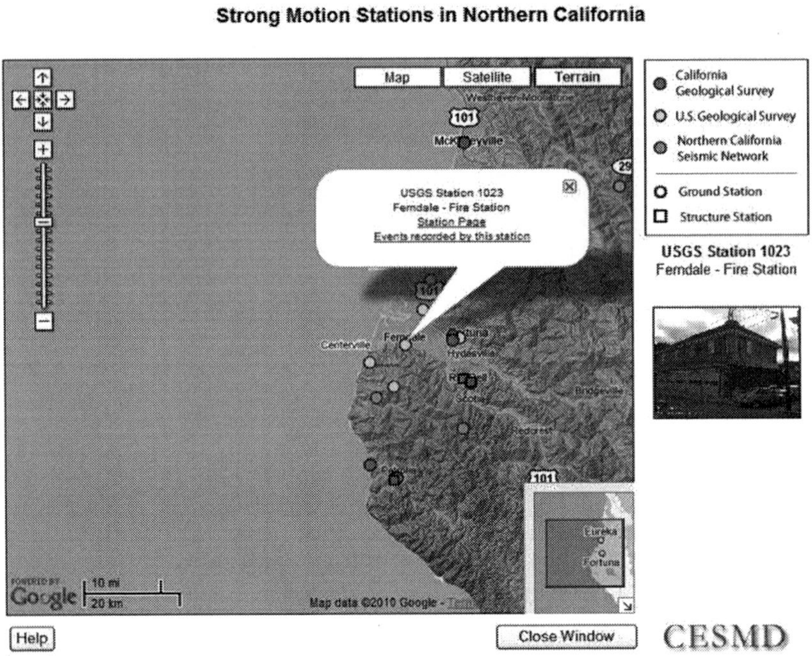

Figure A-12 Northern California strong motion stations map (Source: The Center for Engineering Strong Motion Data operated by CSMIP)

STRUCTURAL IDENTIFICATION OF CONSTRUCTED SYSTEMS   137

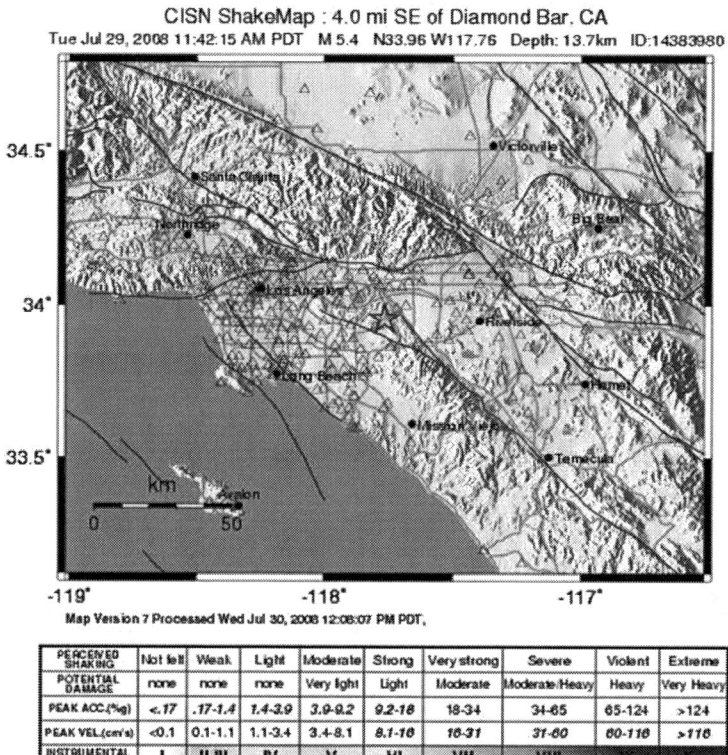

Figure A-13 ShakeMap (Source: The California Integrated Seismic Network operated by CSMIP)

In addition to seismic data, users are provided with a photograph of the station, sensor layouts, information regarding station coordinates, site geology, number of stories above/below ground, plan shape, base and typical floor dimensions, design and instrumentation dates, vertical load carrying and lateral force resisting systems, and foundation type. Such information can give a better insight into the behavior of the structure during ground shaking, and the quality of records.

A.3.6   Interpretation

Despite the large number of structural strong-motion records that have been obtained worldwide and their importance to the earthquake engineering profession, the analysis of these data has not become routine and only a small fraction of the available data has been published, mostly as a result of research studies. Most of these analyses perform modal identification, where parameters of the lower modes of vibration are identified through the analysis of recorded responses (outputs) and ground accelerations (inputs).

Table A-7 presents the structural identification results for the three-story concrete building. For the east-west (lateral) direction, due to the lack of base-rocking measurements, only the flexible base and pseudo-flexible base modal properties can be directly identified. The fixed-base modal properties can be estimated from the other two sets using the methods proposed by Stewart et al. (1998). In estimating the fixed-base properties, Stewart et al. (1998) utilize the relationships proposed by Veletsos and Verbic (1973) who modeled the surrounding soil as a viscoelastic medium. For this indirect identification, values of shear-wave velocity and hysteretic damping ratio for the soil medium are required. Stewart and Stewart (1997) determined these values from de-convolution analyses of the 1989 Loma Prieta earthquake, which were then adopted for the other two earthquakes used in the current study.

Table A-7. Identified frequencies in N-S (transverse) and E-W (lateral) directions

		Fixed Base		Ps.-Flex. Base		Flex. Base	
		NS	EW	NS	EW	NS	EW
Eq.	Mode	[Hz]	[Hz]	[Hz]	[Hz]	[Hz]	[Hz]
LP89	1	6.32	6.07	5.95	5.89	5.44	5.75
LAF07	1	6.65	6.46	6.27	6.09	5.29	5.83
PI07	1	6.33	5.84	5.97	5.57	5.47	5.38

It has been assumed that the structure remains time-invariant (i.e., experiences no significant damage) during all three events. Therefore, it is possible to use the cumulative error method for parameterization of the transfer function. Stewart and Stewart (1997) use a recursive prediction error method to track the time variation of linear system properties during the Loma Prieta earthquake; and the results show no significant time-dependent changes in the first-mode natural frequencies. The magnitudes of vibration during the two other earthquakes are either smaller or in the same range as those for the Loma Prieta earthquake. As such, it is reasonable to assume that the structure remains intact, and its first-mode properties remain time-invariant during these earthquakes, as well.

Figure A-14 displays the variation of all three sets of natural frequencies with the amplitude of vibration in the transverse and lateral directions respectively. As these figures indicate, the natural frequencies corresponding to different conditions of base-fixity range between 5 and 7 Hz. Natural frequencies display some variations with the magnitudes of strong-motion induced vibrations. Stewart et al. (1998) report the uncertainties associated with the natural frequencies identified from the Loma Prieta earthquake to be 0.01~0.02 Hz. Therefore variations of higher order might be associated with factors other than numerical and identification errors.

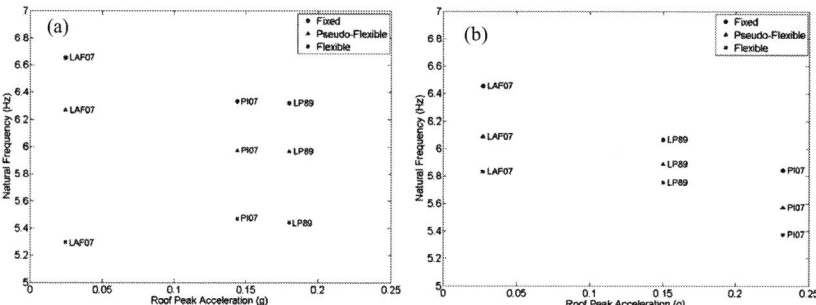

Figure A-14 Variation of (a) North-South (transverse) and (b) East-West (lateral) natural frequencies with vibration amplitude

The observations that are deduced from this case study are as follows:
- In the transverse direction both fixed base and pseudo-flexible base frequencies experience a maximum decrease of approximately 5% (~0.3 Hz) for events with roof peak accelerations larger than 0.1g relative to the event with roof peak acceleration less than 0.05g. On the other hand the flexible base natural frequency rises about 3.5% (~0.2 Hz);
- The differences between frequencies corresponding to different conditions of base-fixity appear to be varying with the shaking intensity. For all events, the difference between the fixed and pseudo-flexible base frequencies is about 0.4 Hz, while the difference between the pseudo-flexible and flexible base frequencies is about 0.5 Hz for the larger events and about 1 Hz for the less severe event. Because the difference between the pseudo-flexible and fixed base frequencies is due to the rocking effects, it may be concluded that the relative rocking effects in the transverse direction are of the same order for all events. On the other hand, because the difference between the flexible and pseudo-flexible base natural frequencies is due to the lateral motions of the foundation with respect to the ground, it may be concluded that the relative effect of lateral movement of the foundation in the transverse direction with respect to the underlying soil is more significant during the smaller event. Furthermore, these effects are more dominant compared to the rocking effects in the transverse direction;
- In the lateral direction, all three sets of natural frequencies drop as the amplitudes of vibration increase in that direction. The maximum amount of decrease varies about 1% for different base-fixity conditions with the largest and smallest differences corresponding to the fixed and flexible base conditions, respectively;
- The effects of lateral motion of the foundation with respect to the ground—i.e., differences between the pseudo-flexible and flexible base

frequencies—are almost within the same range for all three events. On the other hand, the base-rocking effects—i.e., differences between fixed and pseudo-flexible base frequencies—are more significant during the smaller event. Furthermore, during the larger events, the difference between the fixed and pseudo-flexible base frequencies is within the same range as the difference between the pseudo-flexible and flexible base frequencies. However, the former is about twice the latter for the smaller event. Therefore, it can be concluded that during the smaller event in the lateral direction of the specimen structure, the base rocking effects are more significant compared to the effects of lateral motion of the foundation relative to the ground.

The observations above may be interpreted to conclude that (*i*) fixed base frequencies are affected by the level of vibrations more significantly than the flexible base frequencies; (*ii*) in both directions, stronger ground motions lead to an overall decrease in the fixed-base frequencies, which are representative of the flexibility of the structure alone; (*iii*) the increase in the vibration amplitudes leads to an overall increase in the transverse, and an overall decrease in the lateral flexible-base frequencies, which are representative of the flexibility of the structure *and* the surrounding soil media together. Overall, it may be concluded for the specimen structure that the amplitude of vibrations (shaking intensity) appears to influence the structure, and the SSI effects in opposite directions; and thus, the combined effect, which shows up in the flexible-base properties has no discernable overall trend on the observed natural frequencies.

A.4 Guangzhou New TV Tower

A.4.1 Tower Description

The Guangzhou New TV Tower (GNTVT) located in the city of Guangzhou, China, is a supertall tube-in-tube structure with a total height of 610 m as shown in Figure A-15. It comprises a 454 m high main tower and a 156 m high antenna mast. The main tower is composed of a reinforced concrete inner structure and a steel lattice outer structure. The inner structure has a constant ellipse cross-section of 14 m × 17 m throughout the height, while the ellipse cross-section of the outer structure varies with height, being 50 m × 80 m at the ground, 20.65 m × 27.5 m (minimum) at the waist level (280 m high), and 41 m × 55 m at the top (454 m high). The antenna mast supported on the main tower is a steel spatial structure with an octagonal cross-section of 14 m in the maximum diagonal. Designed with functions for sightseeing, TV transmission and cultural entertainment, the GNTVT includes a Ferris wheel, observatory decks, ceremony hall, 4D cinemas, revolving restaurants, and an open-air skywalk.

A.4.2 Objective of St-Id Application

Recognizing the extreme height, unique form and structural complexity of the GNTVT, its safety at both construction and service stages has become the utmost concern of the owner. To ensure safety during construction and operational performance during typhoons and earthquakes, a sophisticated long-term structural health monitoring system consisting of about 800 sensors has been implemented for on-line monitoring of GNTVT at both in-construction and in-service stages. In the meanwhile, a hybrid mass damper (HMD) control system is installed on the main tower while two tuned mass dampers (TMDs) are suspended on the antenna mast for suppressing wind-induced vibration of GNTVT. As the HMD requires the information of current structural condition to make prompt and appropriate action, it is necessary to establish a structural response feedback system to provide thorough information for real-time vibration control. At GNTVT, the SHM system integrates with the vibration control system. It serves as a standby structural response feedback system to ensure reliable and real-time monitoring data can be provided for feedback vibration control. It also has a special function of verifying the vibration control effectiveness because the structural dynamic response both before and after the activation of vibration control system are measured.

Figure A-15 New Guangzhou TV Tower

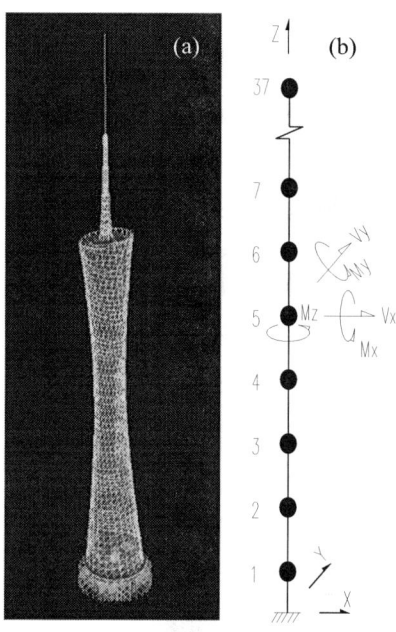

Figure A-16 Finite element models for GNTVT (a) full-order finite element model, (b) reduced-order finite element model

A.4.3 Model Development

A.4.3.1 Full-order finite element model

Using the commercial software SAP2000, a full-order 3D finite element model, as shown in Figure A-16a, has been developed for GNTVT. In this model, the outer structure, antenna mast and connecting beams between the inner and outer structures are simulated by two-node beam elements with six degrees of freedom (DOFs), while the shear-walls of the inner structure and the floor slabs are represented by four-node or three-node shell elements with six DOFs. In total, the model involves 23,714 beam elements, 23,930 shell elements, and 28,305 nodes.

A.4.3.2 Reduced-order finite element model

For damage detection purpose, another reduced-order finite element model has also been formulated for GNTVT (Y. Q. Ni, Xia, Chen et al. 2009). It is a lumped mass model as shown in Figure A-16b. The generalized mass of GNTVT is concentrated in 37 nodes. Each node has two translation DOFs and three rotation DOFs. Linear elastic Euler–Bernoulli beam is employed to model the stiffness of GNTVT. The equivalent stiffness of the beam element is derived from the full-order finite element model using the definition of stiffness. A unit displacement is first applied to one DOF of the beam element in the full-order finite element model. Then, the forces induced by this unit displacement are obtained for all DOFs of the beam element, which forms one column in the stiffness matrix of the beam element. By acting the unit displacement to the rest DOFs, the stiffness matrix of beam element can finally be assembled. MATLAB codes have been programmed to compute the mass and stiffness matrices of the reduced-order finite element model, hence, the modal frequencies and mode shapes.

A.4.4 Experimental Studies

The long-term SHM system for GNTVT has been devised on the basis of a modular design concept which was first practiced in Hong Kong for long-span bridges (Wong and Ni 2009). In accordance with the modular design concept, the SHM system devised for GNTVT consists of six modules, namely, Module 1 – Sensory System (SS), Module 2 – Data Acquisition and Transmission System (DATS), Module 3 – Data Processing and Control System (DPCS), Module 4 – Data Management System (DMS), Module 5 – Structural Health Evaluation System (SHES), and Module 6 – Inspection and Maintenance System (IMS). The SS and DATS are located in the structure, the DPCS, DMS and SHES are inside the monitoring and control room, and the IMS is a portable system.

Figure A-17 shows the deployment of sensors and data acquisition substations on GNTVT. 16 types of sensors are deployed for monitoring three categories of parameters: loading sources (wind, seismic and thermal loading), (ii) structural responses (strain, displacement, inclination, acceleration, and geometric configuration), and (iii) environment effects (temperature, humidity, rain, air pressure, and corrosion). 13 data acquisition sub-stations are employed for in-construction monitoring and six data acquisition sub-stations are utilized for in-service monitoring. Figure A-17 illustrates the deployment of fiber optic sensor on the main tower and antenna mast of the GNTVT. 120 fiber optic sensors have been attached to the outer tube of GNTVT at different heights, and 80 fiber optic sensors are being deployed at the base of the antenna mast (Zhou et al. 2009).

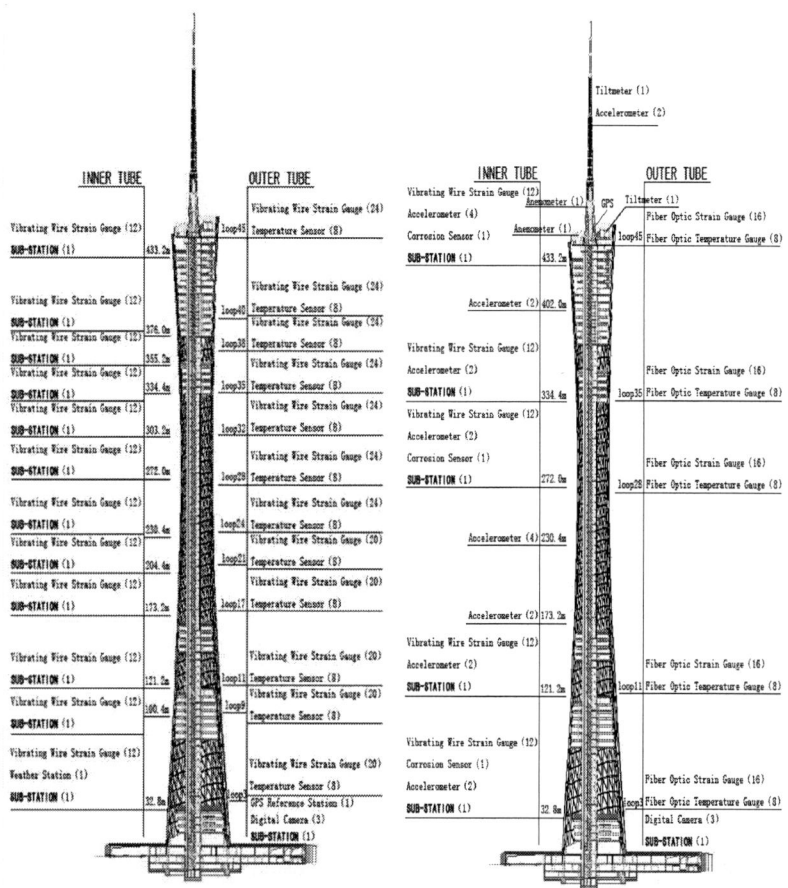

Figure A-17 Deployment of sensors and data acquisition sub-stations on GNTVT: (left) in-construction monitoring and (right) in-service monitoring

Some unique features of the SHM system for GNTVT are (Y. Q. Ni, Xia, Liao et al. 2009):

- In-construction and in-service monitoring combo for life-cycle health track;

144 STRUCTURAL IDENTIFICATION OF CONSTRUCTED SYSTEMS

- Health monitoring and vibration control combo for on-line health monitoring and real-time feedback control;
- Modular system architecture for easy maintenance and upgrade;
- Novel sensors and tailored design customized for special circumstances;
- Hybrid wired and wireless data transmission technology customized for harsh operational conditions;
- User-friendly graphical user interface (GUI) for easy operation;
- Innovative structural health evaluation methodologies beneficial for structural maintenance and management;
- All-round protection customized for severe surrounding environment;
- Web-based data collaboration for remote expert service;
- Edutainment catering for sightseeing and science popularization.

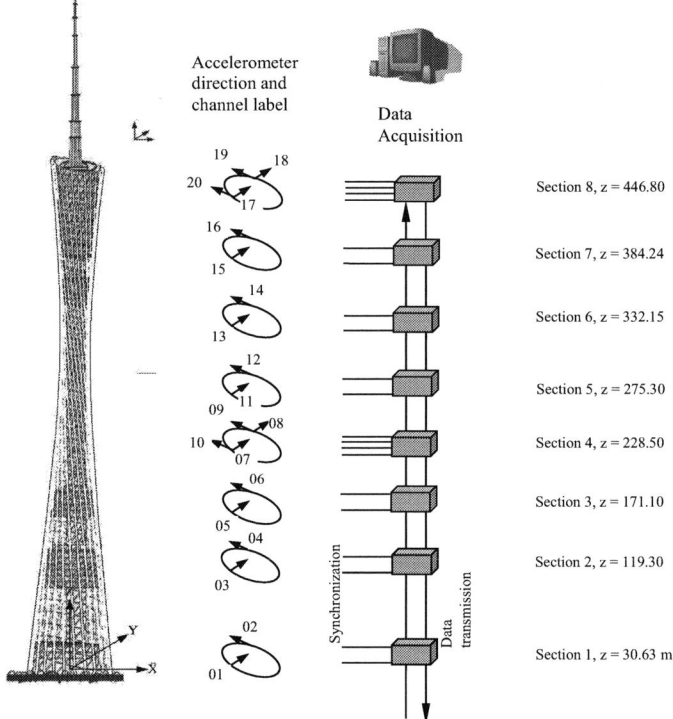

Figure A-18 Layout of accelerometers and data acquisition units for ambient vibration test

STRUCTURAL IDENTIFICATION OF CONSTRUCTED SYSTEMS 145

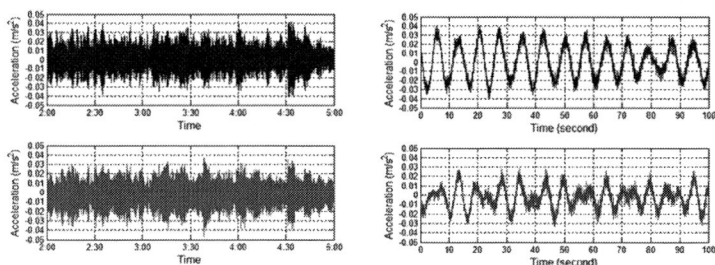

Figure A-19 Example of acceleration response of the main tower

A.4.5 Data Analysis, Model Calibration

Taking the instrumented GNTVT as a test bed (host structure), a SHM benchmark problem for high-rise structures is being developed (Xia et al. 2008). A website has been established for this SHM benchmark problem (Y. Q. Ni and Xia 2010). At present, it is in Phase I, i.e., output-only modal identification and finite element model updating. Two ambient vibration tests have been conducted in the construction stage of GNTVT (Stage 1 and 2). Stage 1 ambient vibration test was performed on 9 and 10 May 2008, when the inner tube was constructed to the height of 402 m (Y. Q. Ni et al. 2008). Stage 2 ambient vibration test was performed in the morning (start from 1:00 am) of 20 March 2009, when the outer tube and inner tube have been completed and the antenna mast was constructed to the height of 460 m (Xia et al. 2009). Both tests lasted two hours. On 8 May 2009, the main structure of GNTVT including the main tower and antenna mast has been completed. By making use of this SHM system, the ambient vibration test at the completed stage is carried out in a continuous and long-term manner. As illustrated in Figure A-18, a total of 20 uni-axial accelerometers were installed at eight levels of the main tower. At the 4th level and 8th level, each section has four uniaxial accelerometers, two for measurement of horizontal vibrations along the long-axis of the inner tube and the other two the short-axis. At the other six levels, each section is equipped with two uniaxial accelerometers, one for the long-axis of the inner tube and the other the short-axis. At each level, a data acquisition unit is placed to collect the acceleration from all sensors within this section. Data are collected from all the 20 accelerometers simultaneously at a sampling frequency of 100 Hz.

Figure A-19 shows the acceleration response of main tower, which were recorded by the SHM system. The modal properties of the main tower identified from the ambient vibration test are summarized in Table A-8. As an illustration, the measured mode shapes for the first two modes are also provided in Figure A-20. Also plotted in this figure are the mode shapes computed by the full-order and reduced-order finite element models. Table A-9 shows the comparison of measured modal frequencies by ambient vibration test with those computed by full-order and reduced-order finite element models. For the full-order finite element model, the difference between computed and measured modal frequency is 1.68% for the 1st mode and the maximum difference is 20.16% for the first eight modes. For the reduced-order finite

146   STRUCTURAL IDENTIFICATION OF CONSTRUCTED SYSTEMS

element model, the difference between computed and measured modal frequency is 1.98% for the 1st mode and the maximum difference is 34.65% for the first eight modes. Upon these measured modal properties, the finite element model of GNTVT is being updated. The validated finite element model will serve as the baseline model for vibration-based damage detection.

Table A-8. Measured modal properties of the main tower.

Mode No.	Frequency (Hz)	Damping ratio (%)	Mode shape
1	0.1012	1.8808	1st short-axis bend
2	0.1478	1.0853	1st long-axis bend
3	0.4763	0.5708	2nd short-axis bend
4	0.5342	0.4207	2nd long-axis bend
5	0.5351	0.3170	1st torsion
6	0.8103	0.3113	3rd short-axis bend
7	0.9801	0.3086	3rd long-axis bend
8	1.2707	0.1798	2nd torsion

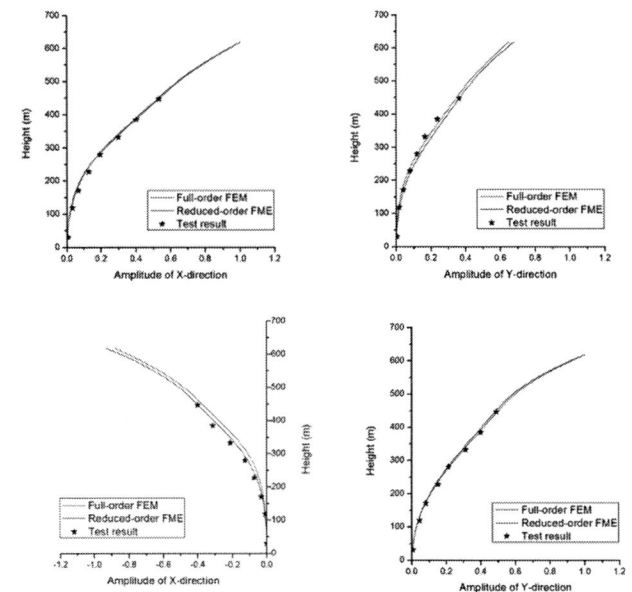

Figure A-20 Mode shapes of the first two modes: (a) 1st short-axis bend, (b) 1st long-axis bend

Table A-9. Comparison of measured and computed modal frequencies.

Mode no.	Frequency (Hz)			Difference (%)	
	Measured	Full-order FEM	Reduced-order FEM	Full-order FEM	Reduced-order FEM
1	0.1012	0.0995	0.1032	-1.68	1.98
2	0.1478	0.1437	0.1522	-2.78	2.98
3	0.4763	0.4419	0.4989	-7.22	4.74
4	0.5342	0.4800	0.6305	-10.15	18.03
5	0.5351	0.4272	0.4519	-20.16	-15.55
6	0.8103	0.6930	1.0654	-14.48	31.48
7	0.9801	0.8543	1.3197	-12.84	34.65
8	1.2707	1.1942	1.1492	-6.02	-9.56

A.4.6 Interpretation and Decision Making

Since a full-scale structure and the real-world measurement data are addressed, the results from this benchmark problem shall be convincing and enable researchers to recognize the obstructions in real life implementations of their damage detection algorithms or techniques. Such a benchmark problem will help reach the most promising directions in future research of SHM, narrow the gap between research and application, and motivate international collaborations in SHM community. After the Phase I task is completed, three more phases will be continued in the first stage of this SHM benchmark problem:

- Damage detection using model-based simulation data. With the full-order and reduced-order finite element models of GNTVT, vibration-based damage detection will be studied using the simulation data. The present study aims to evaluate the applicability and reliability of various damage detection algorithms in the case that (a) only limited modal information is available; (b) there is modeling error; and (c) the measured modal parameters are noise-polluted. The 'measured' modal properties before and after damage will be produced from the full-order finite element model, but only the reduced-order finite element model (in MATLAB format) will be provided to the participants for damage detection to incorporate the modeling error;

- Performance-based optimal sensor placement for structural health monitoring. For large-scale structures such as long-span bridges and high-rise buildings, optimal sensor placement is essential to achieve cost-effective and reliable structural health monitoring. A new method was proposed for optimal placement of sensors on structures for best observing a set of unknown parameters in finite element model updating using Fisher information matrix (Sanayei and Javdekar 2002). The instrumented GNTVT provides a unique paradigm for investigating the optimal sensor placement for damage detection of high-rise slender structures. The existent studies on this topic seek for the sensor locations that can best measure the structural properties. However, a standard SHM system for large-scale civil structures does not pick up the data at the individual measurement points directly; instead, the sensed analog signals are transmitted to the central computer for analog-to-digital data conversion or are first collected by on-structure data acquisition units (sub-stations) for analog-to-digital conversion and then transmitted to the central computer. The data quality can be heavily influenced during signal transmission. As a result, optimal or good locations for structural property monitoring do not imply that the data transmitted to the data acquisition units or central computer are also with high quality. In this task of the benchmark problem, performance-based optimal sensor placement will be studied by considering both structural information aspect and communication/networking constraints. Furthermore, the optimal placement problem can be addressed by exploring multi-scale sensing and data fusion for damage identification; and

- Damage detection using field measurement data. In the latest phase of the benchmark problem, field measurement data of GNTVT before and after 'damage' will be acquired and provided to all interested participants for damage detection study. In this study, the tower immediately before construction of some beams connecting inner tube and outer tube at top will be treated as a 'damaged' structure, and the structure shortly after construction of the top connection beams will serve as the 'intact' state. Another damage scenario will be constructed in which the monitoring data acquired before and after the installation of the 156 m high antenna mast will serve as the field measurement data prior to and posterior to structural damage, respectively. The measured dynamic strain and displacement signals can be used in conjunction with the measured modal properties for structural damage identification.

A.5   Seven-Story RC Building Slice

A.5.1 Building Description

A full-scale seven-story reinforced concrete building slice was tested on the UCSD-NEES shake table in the period October 2005 - January 2006. The test structure represents a slice of a full-scale reinforced concrete building and consists of a main wall (web wall), a back wall (flange wall) perpendicular to the main wall for transversal stability, a concrete slab at each floor level, an auxiliary post-tensioned column to provide torsional stability, and four gravity columns to transfer the weight of the slabs to the shake table. Slotted slab connections are placed between the web and flange walls at floor levels to minimize the moment transfer between the two walls, while allowing the transfer of the in-plane diaphragm forces. Figure A-21 shows the test structure mounted on the shake table.

A.5.2 Objective of Structural Identification Application

The objective of this test program was to verify the seismic performance of a mid-rise reinforced concrete wall building designed for lateral forces obtained from a displacement-based design methodology, which are significantly smaller than those dictated by current force-based seismic design provisions in the United States. As a payload project, system and damage identification studies were performed on the test structure at different damage states to evaluate the performance of the applied methods.

STRUCTURAL IDENTIFICATION OF CONSTRUCTED SYSTEMS   149

Figure A-21 UCSD-NEES test structure (Source: Moaveni et al, 2010; reproduced with permission from Elsevier)

A.5.3 System Identification Methods Applied

Six different state-of-the-art system identification methods, consisting of three input-output and three output-only methods, were applied to dynamic response measurements obtained using DC coupled accelerometers in order to estimate modal parameters (natural frequencies, damping ratios and mode shapes) of the building in its undamaged (baseline) and various damage states. The system identification methods used are: (1) Multiple-reference Natural Excitation Technique in conjunction with Eigen-system Realization Algorithm (MNExT-ERA) (James et al. 1993); (2) Data-driven Stochastic Subspace Identification (SSI) (Van Overchee and de Moore 1996); (3) Enhanced Frequency Domain Decomposition (EFDD) (Brincker et al. 2001); (4) Deterministic-Stochastic Subspace Identification (DSI) (Van Overchee and de Moore 1996); (5) Observer/Kalman filter Identification combined with ERA (OKID-ERA) (Phan et al. 1992); and (6) General Realization Algorithm (GRA) (De Callafon et al. 2008). The first three algorithms are based on output-only data (from white noise and ambient vibration tests) while the latter three are based on measured input and output data (from white noise base excitation tests).

A.5.4 Experimental Studies

A sequence of dynamic tests (68 tests in total) were applied to the test structure during the period October 2005 - January 2006 including ambient vibration tests, free vibration tests, and forced vibration tests (white noise and seismic base excitations) using the UCSD-NEES shake table. The test structure, whose details can be found in Panagiotou et al. (2009), was instrumented with a dense array of DC coupled accelerometers, strain gages, potentiometers, and linear variable displacement transducers, all sampling data simultaneously using a nine-node distributed data acquisition system. The shake table tests were designed to damage the building progressively through four historical earthquake ground motions: (1) longitudinal component of the 1971 San Fernando earthquake (Mw = 6.6) recorded at the Van Nuys station (EQ1), (2) transversal component of the 1971 San Fernando earthquake recorded at the Van Nuys station (EQ2), (3) longitudinal component of the 1994 Northridge earthquake (Mw = 6.7) recorded at the Oxnard Boulevard station in Woodland Hill (EQ3), and (4) 360 degree component of the 1994 Northridge earthquake recorded at the Sylmar station (EQ4). Then, at various levels of damage, ambient vibration tests were performed and low amplitude white noise base excitations were applied through the shake table to the building, which responded as a quasi-linear system with dynamic parameters depending on the level of structural damage. The input white noise base excitation consisted of two 8 minute long realizations of a banded white noise (0.25-25 Hz) process with root-mean-square (RMS) amplitudes of 0.03g and 0.05g, respectively.

In this study, measured response data from 28 longitudinal acceleration channels (three on each floor slab and one on the web wall at mid-height of each story in the direction of base excitation) were used to identify the modal parameters of the test structure. The measured acceleration responses were sampled at 240 Hz resulting in a Nyquist frequency of 120 Hz, which is much higher than the modal frequencies of interest in this study (< 25 Hz).

A.5.5 Data Analysis

Before applying the aforementioned system identification methods to the measured data, all the absolute acceleration time histories were band-pass filtered between 0.5Hz and 25Hz using a high order (1024) Finite Impulse Response (FIR) filter. Furthermore, the absolute horizontal acceleration measurements from the white noise base excitation tests were converted to relative acceleration by subtracting the base/table horizontal acceleration.

Modal parameters of the test structure were identified using the system identification methods defined above based on output-only (for the first three methods) and input-output (for the last three methods) data measured from low amplitude dynamic tests (i.e., ambient vibration tests, 0.03g RMS and 0.05g RMS white noise base excitation tests) performed at various damage states (S0, S1, S2, S3.1, S3.2, and S4). Damage state S0 is defined as the undamaged (baseline) state of the structure before its exposure to the first seismic excitation (EQ1), while damage states S1, S2, S3 and S4 correspond to the state of the structure after exposure to the first (EQ1), second (EQ2), third (EQ3), and fourth (EQ4) seismic excitation, respectively. It should be noted that during damage state S3, the bracing system between the slabs of the test specimen and the post-tensioned column was modified (strengthened and stiffened). Therefore, damage state S3 is subdivided into state S3.1 (before modification of the braces) and state S3.2 (after modification of the braces). Figure A-22 shows the first three longitudinal mode shapes of the test structure identified in its undamaged state. The identified modal parameters are presented and discussed in more detail in Moaveni (2007).

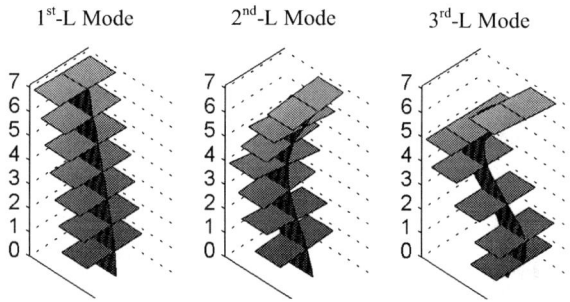

Figure A-22 First three longitudinal (L) mode shapes of the UCSD-NEES test structure in its undamaged state (Source: Moaveni et al, 2010; reproduced with permission from Elsevier)

Figures A-23 and A-24 show the natural frequencies and damping ratios of the first three longitudinal modes identified using three output-only methods based on ambient vibration, 0.03g, and 0.05g RMS white noise base excitation test data. It is observed that: (1) the natural frequencies identified using different methods are reasonably consistent at each damage state considered, while the identified damping ratios exhibit much larger variability. (2) The identified natural frequencies of the three longitudinal vibration modes decrease with increasing level of damage except from damage state S3.1 to S3.2, during which the steel braces were stiffened. The corresponding modal damping ratios do not follow a clear trend as a function of structural damage. (3) At each damage state considered, the identified modal parameters of the first longitudinal mode appear to be the least sensitive to the identification method used, which is most likely due to the predominant contribution of this mode to the total response. (4) The (effective) natural frequency of the first mode identified based on higher amplitude response data is lower than its counterparts identified based on lower amplitude response data at all damage states considered (an average of 20% reduction in the first modal frequencies from ambient vibration data to 0.03g white noise base excitation data and 26% reduction from ambient vibration data to 0.05g white noise base excitation data).

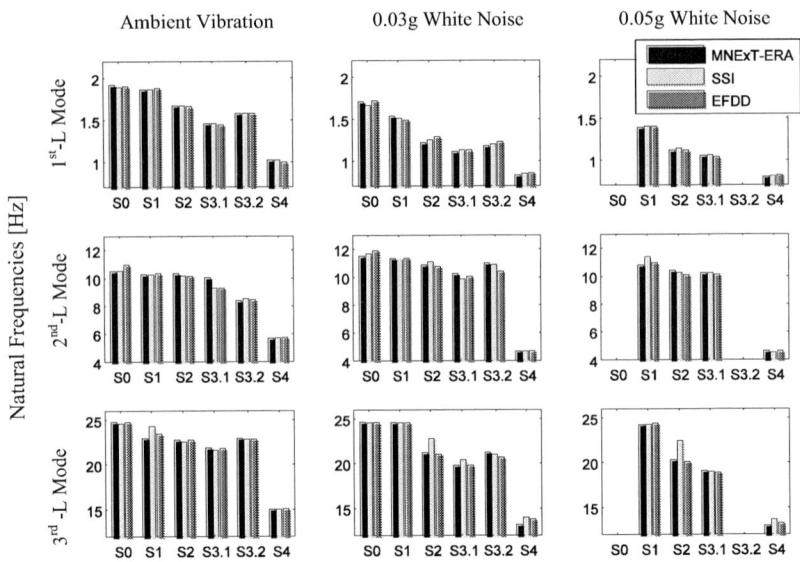

Figure A-23 Natural frequencies of the first three longitudinal modes of UCSD-NEES test structure identified based on ambient and white noise (0.03g, and 0.05g RMS) test data using three output-only methods (Source: Moaveni et al, 2010; reproduced with permission from Elsevier)

This is most likely due to the fact that the test structure is nonlinear (even at the relatively low levels of excitation considered in this system identification study) with effective modal parameters depending strongly on the amplitude of the excitation and therefore of the structural response. (5) In general, larger damping ratios are identified for the three longitudinal vibration modes during the higher amplitude base excitation tests (an average increase of 1.4% in damping ratios from ambient vibration to 0.03g white noise base excitation and 3.1% from ambient vibration to 0.05g white noise base excitation). This is due to the fact that the additional hysteretic damping at higher level of response nonlinearity is included in the equivalent viscous damping model identified using linear system identification methods. The system identification results are then used for damage identification of the building in various damaged states (Moaveni et al. 2010).

STRUCTURAL IDENTIFICATION OF CONSTRUCTED SYSTEMS   153

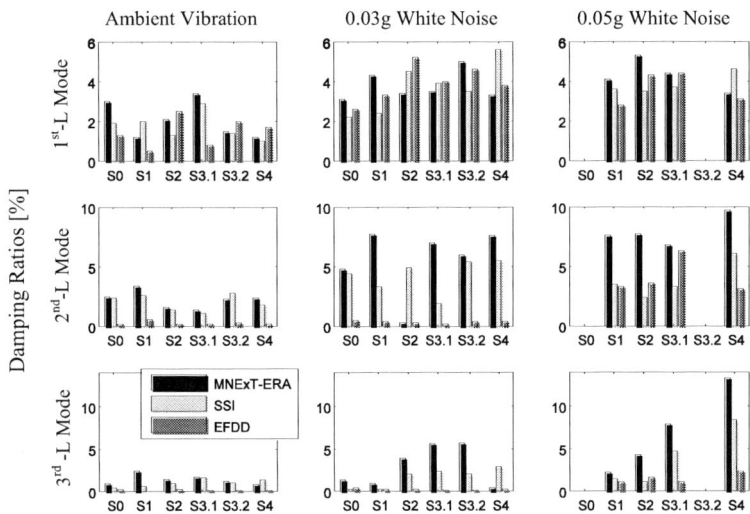

Figure A-24 Damping ratios of the first three longitudinal modes identified based on ambient and white noise (0.03g, and 0.05g RMS) test data using three output-only methods (Source: Moaveni et al, 2010; reproduced with permission from Elsevier)

A.5.6 Damage Identification through Finite Element Model Updating

A sensitivity-based finite element (FE) model updating strategy is applied for vibration based damage identification of this test structure. Three cases of damage identifications are considered based on different sets of modal parameters identified using (I) ambient vibration test data, (II) 0.03g RMS white noise base excitation test data, and (III) 0.05g RMS white noise base excitation test data. In each case of damage identification, once a reference model is obtained, 10 updating parameters (corresponding to 10 substructures) are updated from the reference FE model (at the undamaged/baseline state S0) to states S1, S2, S3.1, and S4. These 10 substructures represent the web wall, 6 along the first three stories (two per story) and 4 along the higher stories (one per story). The values of the stiffness parameters of the remaining substructures are kept fixed at the corresponding values in the reference FE model. For each of the considered states of the building, the natural frequencies and mode shapes of the first three longitudinal vibration modes are used in the objective function for damage identification, resulting in a residual vector with 42 components (i.e., 3 natural frequencies and 3 vibration mode shapes with 13 components each).

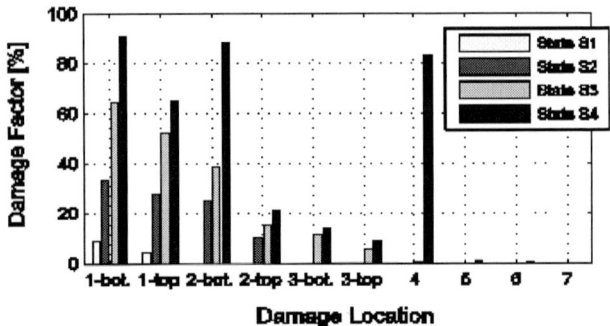

Figure A-25 Identified damage factors at various substructures for damage identification case I (based on ambient vibration data)

The identified damage factors (loss of stiffness in each substructure relative to the reference FE model) obtained at different damage states for Case I (based on ambient vibration data) are presented in a bar plot in Figure A-25. These results indicate that: (1) the severity of structural damage increases as the building is exposed to stronger earthquake excitations; and (2) the extent of damage decreases rapidly along the height of the building (damage concentrated in the two bottom stories), except for a false alarm in the fourth story at state S4. This large identified damage factor in the fourth story may be due to the facts that the updating parameters account for the effect of damage in other non-updating elements such as the floor slabs, flange wall, post-tensioned column and the connections between floor slabs and post-tensioned column, and the identified damage factors at state S4 are in general characterized by a higher level of estimation uncertainty than at the previous (lower) damage states.

Overall, the damage identification results obtained for the three cases do not match exactly, but they are consistent with the actual damage observed in the building which shows a concentration of damage at the bottom two stories of the web wall. Pictures of the actual damage at the bottom two stories of the web wall at state S4 are shown in Figure A-26. The difference in the identified damage results across the three cases considered is mainly due to the significant difference in the identified modal parameters used in the three cases. The assumption of a quasi-linear dynamic system is progressively violated with increasing level of excitation. Therefore, the identified modal parameters (especially of the first mode) corresponding to different levels of excitation are significantly different. It is worth noting that the ambient vibration data satisfy better the assumption of system linearity and therefore are more appropriate as input for linear FE model updating. This is confirmed by the fact that the analytical modal parameters obtained from the updated FE models are in better agreement with their experimentally identified counterparts in Case I than in Cases II and III. The damage identification results are presented and discussed in more detail in Moaveni et al. (2010).

STRUCTURAL IDENTIFICATION OF CONSTRUCTED SYSTEMS

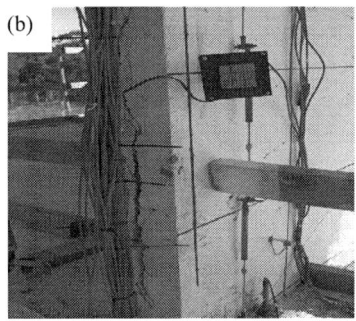

Figure A-26 (a) Crack opening at the bottom corner of the first story of the web wall during EQ4 (at instant of time near maximum base rotation) and (b) splitting crack due to lap-splice failure at the bottom of the second story after EQ4

A.6   Building Substructure Example: Composite Structural Floor System

A.6.1   Structure Description

The structure is the second floor (one story height above ground) of a purpose-built office building in Leeds, UK. The structure has steel primary beams at 6 m centres, secondary beams at 3 m centres and steel columns approximately on a 6 m ×12 m grid. Composite steel-concrete slabs span between the secondary beams. The floor is based loosely on a regular set of bays totalling 72 m ×24 m with additional voids for staircases. Though details within close proximity to voids and geometric irregularities differ, in general the primary cellular beams are constructed from an upper Tee 457×191×89UB and a lower Tee 610 × 229 × 113UB with voids of diameter 550 mm at 750 mm centres. Secondary beams are 254×146×31UB and the columns are 254×254×73UC. Photographs taken on-site have provided an estimate for the concrete slab as being 130 mm deep, with 60 mm trapezoidal decking.

Figure A-27 upper shows the floor layout (dimensions in mm), lower right shows the office environment (and the columns), lower right the primary and secondary beams supporting trapezoidal steel decking.

A.6.2 Objective of St-Id Application

The floor is considered by occupants (employees of a British structural engineering consulting firm) to be quite lively, but not sufficiently lively to attract complaints. The floor was chosen as the test bed for a new active vibration control system (Diaz and Reynolds 2010) requiring creation of a modal model for simulations, evaluation of structural contributions to dynamic performance through finite element modeling and correlation and measurements of vibration response to walking with and without active control in operation.

156 STRUCTURAL IDENTIFICATION OF CONSTRUCTED SYSTEMS

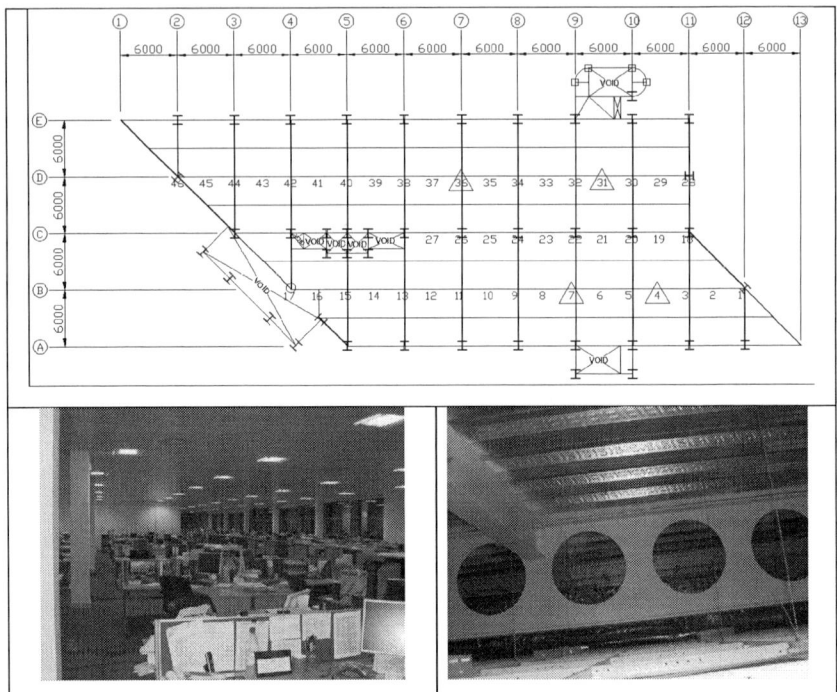

Figure A-27: Upper, structural system; lower left office environment showing columns; right view below floor showing primary and secondary beams and trapezoidal steel decking for RC slab

A.6.3 Model Development

The ANSYS commercial FE software was used to model the floor. The composite slabs were modelled using orthotropic SHELL63 elements, where the slab thickness and density were constant throughout but the orthotropic behaviour of the slab in directions of the primary and secondary beams (due to the trapezoidal steel decking see in Figure A-27) were modelled by reduced Young's modulus (nominally 38MPa) in the secondary beam direction. The primary and secondary beams were modelled using BEAM44 elements which allow for taper and centroidal offsets. Composite connections between the beams and slabs were modelled using offset centroids of the beams and slab. Columns were modelled (without offsets) using the relatively simple BEAM4 elements. Both BEAM44 and BEAM4 elements incorporate tension, compression, bending and torsion capabilities.

The columns were assumed to be fixed one storey above and below the floor under consideration. All other internal connections were assumed to be fixed, an assumption generally taken as valid because the very small deflections resulting from walking-induced vibrations are not sufficient to cause significant rotation at joints, even if those joints are designed to be pinned with regards to ultimate limit state analysis. Imposed loads and non-structural dead loads were modelled as additional mass on the slab elements. Figure A-28 shows the initial FE model (slab is not shown for clarity).

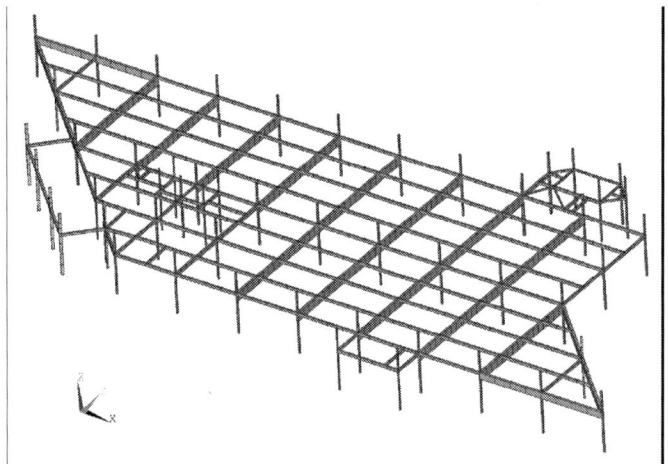

Figure A-28: Beams and columns of initial FE model (slab not shown for clarity)

A.6.4 Experimental Studies

Test Point (measurement) locations for the modal test are shown Figure A-29, with vertical accelerometers located at column and mid-bay locations wherever possible. Attention was paid to TP04 and its surroundings because it was perceived to be a particularly lively location on the floor. Because the vibration perception was particularly acute at this point, this was a good initial candidate for the installation of the shaker for the subsequent active vibration control studies.

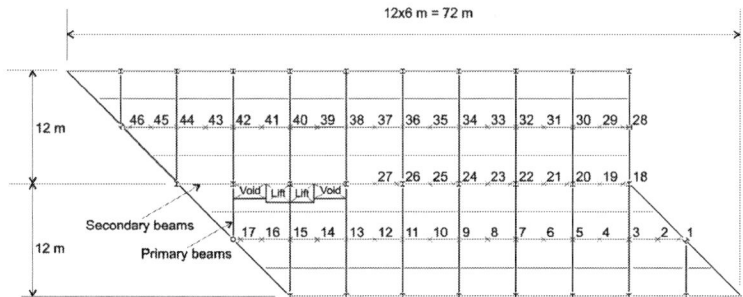

Figure A-29: Test grid

Modal testing was carried out using artificial excitation supplied by two APS Dynamics Model 400 electrodynamic shakers operated in inertial mode. Four excitation points were used (TPs 04, 07, 31 and 36) and responses were measured at all TPs, resulting in 4 columns of the frequency response function (FRF) matrix. The modal testing was carried out using continuous uncorrelated random excitation with two excitation points at a time (i.e. multi-input multi-output or MIMO modal testing). Time domain data blocks were of duration 20 s giving a frequency resolution of 0.05 Hz. The number of averages was 80 with 75% overlapping and a Hanning window was applied to all data blocks.

The magnitudes of the driving point mobility FRFs acquired are shown in Figure A-30 where force and response are measured at the same point. From a visual inspection, there are approximately nine modes between 4 and 10 Hz. The lowest mode occurs at 4.86 Hz and the highest peak occurs at TP04 at approximately 6.4 Hz. TP04 is the point on the structure where the response was subjectively assessed to be highest.

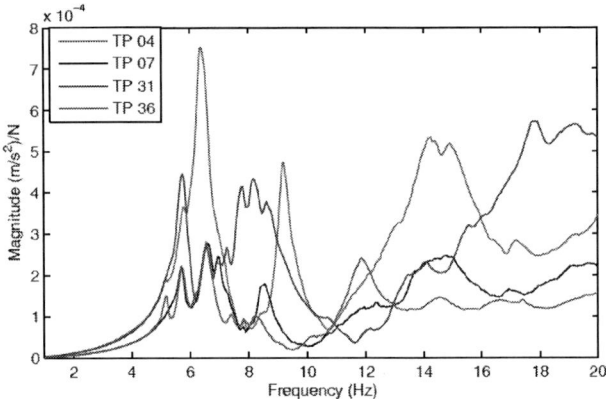

Figure A-30: Driving point mobilities (accelerance or inertance values)

On-site modal parameter estimation was carried out on the full set of acquired FRFs using the ME'scope suite of software. In particular, mode indicator functions were first calculated to give an indication of the locations of vibration modes and then the multiple reference orthogonal polynomial algorithm was used to estimate the modal properties, including modal mass for mode shape scaling. Between 4.86 and 9.19 Hz, 13 modes were estimated. Fig. A-31 shows the estimated vibration modes, which were dominant at TP04. The vibration mode at approximately 6.37 Hz is the most likely to be excited by pedestrian excitation; this mode has a damping ratio of 3% and a modal mass of approximately 20 tonnes.

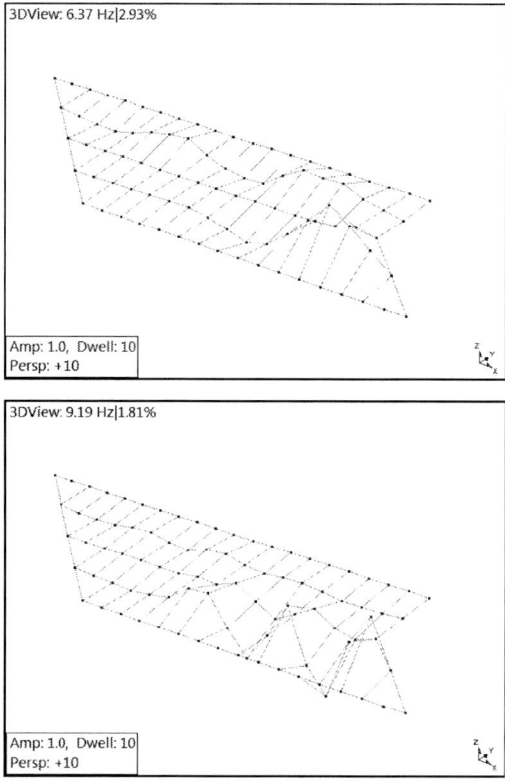

Figure A-31: Dominant vibration modes at TP04 with frequency and damping estimates

A.6.5 Data Analysis, Model Calibration

The primary aim of the experimental modal analysis (EMA) was to generate an experimental modal model for designing and simulating the performance of the active vibration control system. Such a model represents reality in operational conditions and is chosen for performance simulations wherever possible and with access to the full-scale structure. For a-priori simulations only finite element analysis (FEA) is available and modelling technology for floors engages a different set of uncertainties. Both FEA and EMA can produce modal models that are suitable for performance simulations for assessment of vibration serviceability.

160 STRUCTURAL IDENTIFICATION OF CONSTRUCTED SYSTEMS

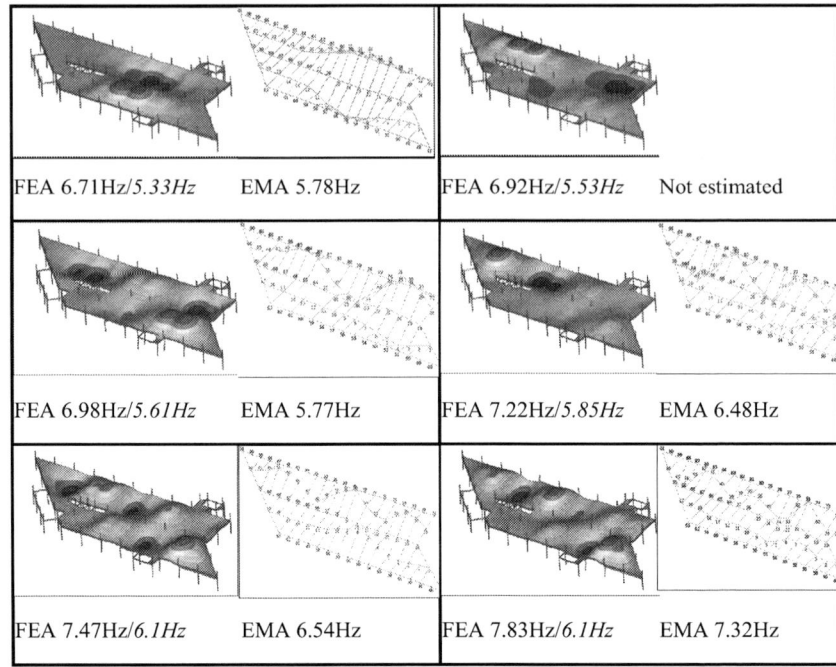

FEA 6.71Hz/*5.33Hz*	EMA 5.78Hz	FEA 6.92Hz/*5.53Hz*	Not estimated
FEA 6.98Hz/*5.61Hz*	EMA 5.77Hz	FEA 7.22Hz/*5.85Hz*	EMA 6.48Hz
FEA 7.47Hz/*6.1Hz*	EMA 6.54Hz	FEA 7.83Hz/*6.1Hz*	EMA 7.32Hz

Figure A-32: Matching of first six preliminary FEA modes and corresponding EMA modes, with updated FEA frequencies in italics

For this floor limited model calibration was undertaken in order to improve understanding of the performance of the structural system. Since this type of flooring system is common in the UK, such a correlation study has benefits for a-priori analysis of similar structures that may be problematic. Figure A-32 shows matching of selected FEA and EMA modes, not necessarily the same as the critical modes for the AVC study, but intended for manual updating. An independent modal analysis was performed using a different mode estimation technique, explaining the slight difference in frequencies to the EMA results presented in Figure A-31.

Figure A-32 shows a reasonable correlation between the FE and EMA shapes for the first six modes, with the exception of the second FE mode which was not picked up by the EMA study. One area of uncertainty is the additional stiffness and additional mass from non-structural elements such as storage areas and office equipment. The natural frequencies from the preliminary FE study are noticeably higher than those from the EMA study, indicating a lack of mass or excessive stiffness in the model which could derive from differences in slab depth or effects of non-structural components. Other possibilities are incorrect assumption about concrete modulus and the degree of composite action.

Increasing the slab depth causes lower modes to decrease in frequency (because these modes are global with concrete behaving more as added mass) while higher modes

STRUCTURAL IDENTIFICATION OF CONSTRUCTED SYSTEMS 161

increase in frequency (because the stiffness of the slab dominates with more local bending in higher modes). Factors such as material modulus and member geometric properties could not be in error enough to explain the differences so possible reasons for a lack of composite action were explored. Adjusting shear lag in the slab and cracking in the concrete above hogging regions resulted in insignificant changes in natural frequencies. So, some mechanism exists in the real structure through which stiffness is lost. The best improvement was obtained by a change in the offset for the beams and a small reduction in the Young's Modulus of the concrete. The updated FEA model frequencies are given in italics in Figure A-32.

Figure A-33 shows a comparison of the FRF obtained from EMA with that from updated FEA for TP04, the location of most lively response, assuming a damping value of 2.5% in the FEA in line with average of values from the modal test. The important features of the EMA in the frequency range of concern are re-created acceptably by the FEA.

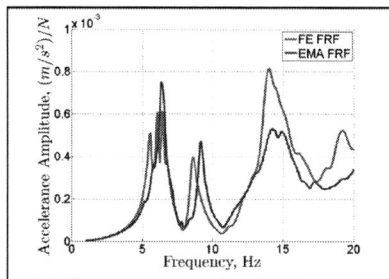
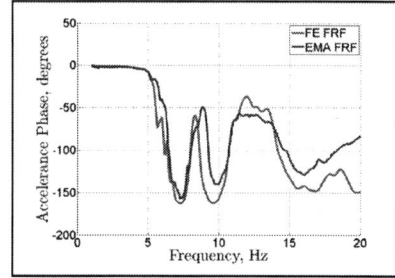

Figure A-33: Matching of EMA and FEA frequency response functions for TP04, using updated FEA model

A.6.6 Interpretation and Decision Making

For this type of structure the major concern is with vibration serviceability due to footfall-induced vibrations. Both FEA and EMA results can be used for performance simulations using either published design guidance (Pavic and Willford 2005), referred to as CSTR43, or by direct simulation using measured ground reaction force (GRF) time histories as moving dynamic loads (Brownjohn et. al 2004).

For a prototype structure, whose design may have been adjusted on the basis of such simulations using a-priori FEA, walking response measurements can be made as the final proof test of actable performance.

Figure A-34 shows simulations using the updated FE model showing the response hotspot around TP04 for walking at 1.6Hz, exciting response in modes around 6.5Hz by the fourth harmonic of the walking force fundamental frequency component. The simulations use first principles approaches of CSTR43 implemented using bespoke MATLAB software VSATs (Pavic et. al. 2010). The numerical values are 'R factors' referenced to a RMS acceleration value of 0.005m/sec2 calculated with a 1 second averaging time and with ISO-standard frequency weighting.

162 STRUCTURAL IDENTIFICATION OF CONSTRUCTED SYSTEMS

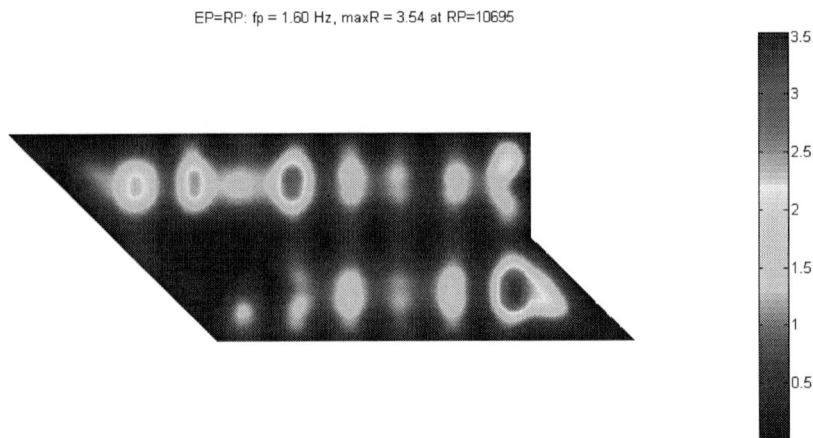

Figure A-34: Vibration response simulations using first-principles approaches and published design guidance on footfall forces. R=1 denotes frequency weighted RMS response of $0.005/sec^2$

The structure is classed as a 'low frequency floor' because its performance with respect to footfall forces generated is dominated by modes in which resonance can be generated by strong components of the walking force occurring at lower multiples of the pacing rate. High-performance (i.e. low response) floors typically found in hospitals and micro-chip plants are classed as 'high frequency floors' since their response is dominated by rapid transient decay of modes with frequencies above 10Hz due to the impulse-like force characteristic of individual footfalls.

Vibration tests as described in this study are often required to demonstrate compliance of an as-built structure with design specifications (i.e. a maximum R-factor, according to usage), while a-priori modeling, influenced by experience of model/test correlations of similar structures seeks to use best practice to predict performance capability before construction, giving an opportunity to adjust a poor design.

The maximum R-value for the floor (over all pacing rates and response points) is 7.3. This is just within acceptance limits for an office floor.

With the main objective for this particular study being the development of an active vibration control, the outcome of the experimental study has been generation of an appropriate modal model for design of the AVC. Figure A-35 shows on-site evaluation of the AVC designed using the EMA results. AVC performance was assessed for controlled excitation, driving with one shaker and controlling with another, and for more usual (design) scenario of footfall loads due to a single pedestrian.

STRUCTURAL IDENTIFICATION OF CONSTRUCTED SYSTEMS   163

Figure A-35: Active vibration control system in operation: (left) simulating excitation with one shaker while controlling with another and (right) in action reducing floor vibrations induced by walking

Figure A-36 shows the success of the AVC in controlling response at the most lively point, TP04. The figure also shows the in-situ measured response to walking. The red lines are the RMS envelope, and for the uncontrolled floor, the values are a good match the predictions of Figure A-34. The exercise demonstrates the capability of AVC system for significant improvement in floor vibration performance

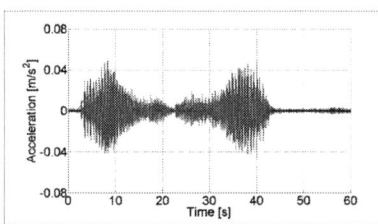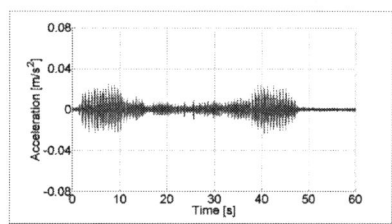

Figure A-36: Floor response for walking at 1.58Hz (left) without AVC and (right) with AVC engaged

A.7   References

Bentz, A., and Kijewski-Correa, T. (2008). "Predictive models for damping in buildings: The role of structural system characteristics." *Structures Congress: 18th Analysis and Computation Specialty Conference,* Vancouver, Canada.

Bentz, A., Young, B., Kijewski-Correa, T., and Abdelrazaq, A. (2010). "Finite element modeling of common lateral systems in tall buildings: Insights from full-scale monitoring." *Structures Congress 2010,* ASCE, Orlando, FL.

Brincker, R., Ventura, C., and Andersen, P. (2001). "Damping estimation by frequency domain decomposition." *Proceedings of International Modal Analysis Conference (IMAC XIX)*, Kissimmee, FL.

Brownjohn, J. M. W., Pan, T. C., and Cheong, H. K. (1998). "Dynamic response of republic plaza, Singapore." *Structural Engineer*, 76, 221-226.

Brownjohn, J. M. W., Pavic, A. and Omenzetter, P. (2004) A Spectral Density Approach for Modelling Continuous Vertical Forces on Pedestrian Structures Due to Walking. *Canadian Journal of Civil Engineering*, Vol. 31, No. 1, pp. 65-77

California Geological Survey. (2007). *California geological survey - about CSMIP*, <http://www.conservation.ca.gov/cgs/smip> (January 4, 2011).

CESMD. (2011). *Center for engineering strong motion data*, <http://strongmotioncenter.org> (January 4, 2011).

Diaz, I.M. and Reynolds, P. (2010) Acceleration feedback control of human-induced floor vibrations. *Engineering Structures*, Vol. 32, No. 1, January, pp. 163-173

De Callafon, R. A., Moaveni, B., Conte, J. P., He, X., and Udd, E. (2008). "General realization algorithm for modal identification of linear dynamic systems." *Journal of Engineering Mechanics*, 134(9), 712-722.

Huang, M. J. (2006). "Utilization of strong-motion records for post-earthquake damage assessment of buildings." *International Workshop on Structural Health Monitoring and Damage Assessment*, Taichung, Taiwan.

Huang, M. J., and Shakal, A. F. (2001). "Structure instrumentation in the California strong motion instrumentation program." *Strong motion instrumentation for civil engineering structures*, M. Erdik, M. Celebi, V. Mihailov and N. Apaydin, eds., Kluwer, Dordrecht, Netherlands, 17-31.

James, G. H., Carne, T. G., and Lauffer, J. P. (1993). *The natural excitation technique for modal parameter extraction from operating wind turbines*, SAND92-1666, UC-261, Sandia National Laboratories, Albuquerque, NM.

Kareem, A., Kijewski, T., and Tamura, Y. (1999). "Mitigation of motion of tall buildings with recent applications." *Wind Structures*, 2(3), 201-251.

Kijewski-Correa, T., and Kareem, A. (2007). "Monitoring serviceability limit states in civil infrastructure: Lessons learned from the Chicago full-scale monitoring experience." *Proceedings of 6th International Workshop on Structural Health Monitoring*, Palo Alto, CA.

Kijewski-Correa, T., Kilpatrick, J., Kareem, A., Kwon, D. K., Bashor, B., Kochly, M., Young, B. S., Abdelrazaq, A., Galsworthy, J. K., Isyumov, N., Morrish, D., Sinn, R. C., and Baker, W. F. (2006). "Validating wind-induced response of tall buildings: Synopsis of the Chicago full-scale monitoring program." *ASCE Journal of Structural Engineering,* 132(10), 1509-1523.

Kijewski-Correa, T., and Kochly, M. (2007). "Monitoring the wind-induced response of tall buildings: GPS performance and the issue of multipath effects." *Journal of Wind Engineering and Industrial Aerodynamics,* 95(9-11), 1176-1198.

Kijewski-Correa, T., and Pirnia, J. D. (2007). "Dynamic behavior of tall buildings under wind: Insights from full-scale monitoring." *Structural Design of Tall and Special Buildings,* 16, 471-486.

Kijewski-Correa, T., and Cycon, J. (2007). "System identification of constructed buildings: Current state-of-the-art and future directions." *Structural Health Monitoring and Intelligent Infrastructure,* Vancouver, Canada.

Kijewski-Correa, T., Taciroglu, E., and Beck, J. L. (2008). "System identification of constructed facilities: Challenges and opportunities across hazards." *Structures Congress,* Vancouver, Canada.

Kijewski-Correa, T., Young, B., Baker, W. F., Sinn, R., Abdelrazaq, A., Isyumov, N., and Kareem, A. (2005). "Full-scale validation of finite element models for tall buildings." *CTBUH 7th World Congress,* New York.

Lin, C. C., Wang, C. E., and Wang, J. F. (2003). "On-line building damage assessment based on earthquake records." *Structural Health Monitoring and Intelligent Infrastructure,* Tokyo, Japan.

Moaveni, B. (2007). "System and damage identification of civil structures." Department of Structural Engineering, UCSD, San Diego, CA.

Moaveni, B., He, X., Conte, J. P., and Restrepo, J. I. (2010). "Damage identification study of a seven-story full-scale building slice tested on the UCSD-NEES shake table." *Journal of Structural Safety,* 32(5), 347-356.

NEES@UCLA. (2010). *Field vibration testing and analytical studies of a four-story RC building,* <http://www.nees.ucla.edu/fourseasons.html> (January 4, 2011).

Ni, Y. Q., & Xia, Y. (2010). *A benchmark problem for structural health monitoring of high-rise slender structures,* <http://www.cse.polyu.edu.hk/benchmark/> (January 4, 2011).

Ni, Y. Q., Xia, Y., Chen, W. H., Ko, J. M., and Liao, W. Y. (2009). "ANCRiSST benchmark problem on structural health monitoring of high-rise slender structures-phase I: Reduced-order finite element model." *5th International Workshop on Advanced Smart Materials and Smart Structures Technology,* Boston, MA.

Ni, Y. Q., Xia, Y., Liao, W. Y., and Ko, J. M. (2009). "Technology innovation in developing the structural health monitoring system for Guangzhou new TV tower." *Structural Control and Health Monitoring,* 16(1), 73-98.

Ni, Y. Q., Xia, Y., Lu, Z. R., and Tam, H. Y. (2008). "Technological issues in developing structural health monitoring for a supertall structure." *3rd World Congress on Engineering Asset Management and Intelligent Maintenance Systems,* Beijing, China.

Panagiotou, M., Restrepo, J. I., and Conte, J. P. (2009). "Shake table test of a 7-story full scale building slice - phase I: Rectangular wall." *Journal of Structural Engineering,* in review.

Pavic, A., Brownjohn, J.M.W. and Zivanovic S. VSATs software for assessing and visualizing vibration serviceability based on first principles. ASCE Congress 2010, Orlando, Florida, USA, 12-15 May 2010.

Pavic, A. and Willford, M. 2005. Vibration Serviceability of Post-Tensioned Concrete Floors. Appendix G in Post-Tensioned Concrete Floors Design Handbook, 2nd Edition. Technical Report 43. Concrete Society. Slough, UK.

Phan, M., Horta, L. G., Juang, J. -., and Longman, R. W. (1992). *Identification of linear systems by an asymptotically stable observer,* 3164, NASA, Washington, DC.

Safak, E. (1991). "Identification of linear structures using discrete-time filters." *Journal of Structural Engineering,* 117(10), 3064-3085.

Sanayei, M. and Javdekar C. N., "Sensor Placement for Parameter Estimation of Structures using Fisher Information Matrix," Proceedings of 7th International Conference on Applications of Advanced Technology in Transportation (AATT 2002), Cambridge, MA, August 5-7 2002, pp. 386-393.

Shakal, A. F., and Huang, M. J. (1985). *Standard tape format for CSMIP strong-motion data tapes,* OSMS 85-03, California Department of Conservation, Division of Mines and Geology, Office of Strong Motion Studies, Sacramento, CA.

Shakal, A. F., Huang, M. J., and Graizer, V. M. (2004). "CSMIP strong-motion data processing." *COSMOS Invited Workshop on Strong-Motion Record Processing,* Richmond, CA.

Skolnik, D., Lei, Y., Yu, E., and Wallace, J. (2006). "Identification, model updating, and response prediction of an instrumented 15-story steel-frame building." *Earthquake Spectra,* 22(3), 781-802.

Stewart, J. P., and Fenves, G. (1998). "System identification for evaluating soil-structure interaction effects in buildings from strong motion recordings." *Earthquake Engineering and Structural Dynamics*, 27(8), 869-885.

Stewart, J. P., Seed, R. B., and Fenves, G. (1998). *Empirical evaluation of inertial soil-structure interaction effects,* Pacific Earthquake Engineering Research Center, University of California, Berkeley, Berkeley, CA.

Stewart, J. P., and Stewart, A. F. (1997). *Analysis of soil-structure interaction effects on building response from earthquake strong motion recordings at 58 sites,* Earthquake Engineering Research Center, University of California, Berkeley, Berkeley, CA.

Van Overchee, P., and de Moore, B. (1996). *Subspace identification for linear systems: Theory, implementation and applications,* Kluwer Academic Publishers, Dordrecht, Netherlands.

Van Overschee, P., and DeMoor, B. (1994). "N4SID: Subspace algorithms for the identification of combined deterministic-stochastic systems." *Automatica,* 30(1), 75-93.

Veletsos, A. S., and Verbic, B. (1973). "Vibration of viscoelastic foundations." *Earthquake Engineering and Structural Dynamics,* 2, 87-102.

Wong, K. Y., and Ni, Y. Q. (2009). "Modular architecture of structural health monitoring system." *Encyclopedia of structural health monitoring,* C. Boller, F. K. Chang and Y. Fujino, eds., John Wiley & Sons, Chichester, UK, 2009-2015.

Xia, Y., Ni, Y. Q., Ko, J. M., and Chen, H. B. (2008). "ANCRiSST benchmark problem on structural health monitoring of high-rise slender structures." *Proceedings of the 4th International Workshop on Advanced Smart Materials and Smart Structures Technologies,* Tokyo, Japan.

Xia, Y., Ni, Y. Q., Ko, J. M., Liao, W. Y., and Chen, W. H. (2009). "ANCRiSST benchmark problem on structural health monitoring of high-rise slender structures - phase I: Field vibration measurement." *Proceedings of the 5th International Workshop on Advanced Smart Materials and Smart Structures Technology,* Boston, MA.

Yu, E., Taciroglu, E., and Wallace, J. W. (2007). "Parameter identification of framed structures using an improved finite element model updating method, part I: Formulation and verification." *Earthquake Engineering and Structural Dynamics,* 36, 619-639.

Yu, E., Wallace, J. W., and Taciroglu, E. (2007). "Parameter identification of framed structures using an improved finite element model updating method, part II:

Application to experimental data." *Earthquake Engineering and Structural Dynamics*, 36, 641-660.

Zhou, H. F., Ni, Y. Q., Liao, W. Y., Tam, H. Y., and Liu, S. W. (2009). "Structural health monitoring system for steel antenna mast of Guangzhou television and sightseeing tower." *Sixth International Conference on Advances in Steel Structures*, Hong Kong.

Appendix B
Case Studies on the Structural Identification of Bridges

B.1 Henry Hudson Bridge

B.1.1 Bridge Description

The Henry Hudson Bridge is a major long-span steel arch structure that serves as a key river-crossing for New York City (Figure B-1). The bridge features two deck levels supported by two, 256 m long plate girder arches that provide a vertical clearance of 44 m. The arch span is flanked at its ends by two steel tower structures, viaduct spans and approach spans. The viaducts at the northern and southern ends are supported by steel sub-structures at intervals of 18 m. The width of the bridge measured from center-to-center between the vertical columns is 15 m.

B.1.2 Objective of Structural Identification Application (Step 1)

The primary objective for conducting the St-Id of the bridge (Step 1 of the St-Id process) was to support a seismic vulnerability assessment study and, if required, a subsequent retrofit design (Grimmelsman et al. 2007; Pan 2007).

B.1.3 System Identification Method Development (Step 2)

An a priori finite element model of the bridge was developed by Parsons Corporation using the available design and construction documents as well as the maintenance and inspection records for the bridge. This element-level 3D FE model was constructed in SAP2000 (CSI) and incorporated the main arch span and the two viaduct spans. Both the upper and lower level reinforced concreted decks were discretized using shell elements with six degrees of freedom at each node in order to capture in-plane and out-of-plane deformations. The deck stringers were not explicitly simulated, but their contribution to the stiffness of the floor system was smeared into the deck shell elements. Space frame elements were used to represent the floor beams, verticals, arch ribs, and bents of the sub-structure, while the bracing and tower truss members were modeled with truss elements to mimic the actual end connections.

Figure B-1. Henry Hudson Bridge

170   STRUCTURAL IDENTIFICATION OF CONSTRUCTED SYSTEMS

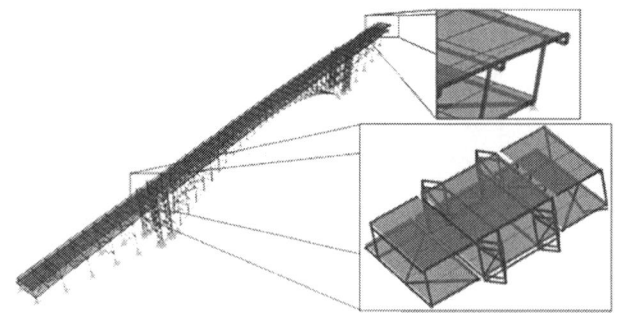

Figure B-2. A-priori finite element model of the test bridge

The main arch, towers, as well as substructure bents on the south and north viaduct spans are all anchored in massive concrete blocks which are founded on rock. The FE model ignored any interactions between the structure and the foundation blocks as well as the inertia effects of the foundations, utilizing pinned or fixed restraints as appropriate at the interface of each of the steel substructure elements and the concrete blocks. The joints located at the interface between the deck and tower at both deck levels incorporated movement systems designed to accommodate longitudinal deformations of the structure under temperature change. These systems were simulated using gaps in the FE model. The three layers in the elevation of the bridge superstructure – the two decks and the arch ribs – were connected with vertical members, and massless rigid links were utilized to maintain the geometry of the connections between the deck and exterior roadway girders and between the deck and floor beams. This model is shown in Figure B-2.

B.1.4   Experimental Studies (Step 3)

The bridge was tested using two sensor layouts (Figure B-3 and Figure B-4) that maintained a sufficiently dense spatial array in which all sensor measurements were collected simultaneously. In the first test stage, a total of 36 accelerometers were installed on the north half of the arch-span, north tower and north viaduct. The south-half of the arch span, south tower and south viaduct were tested in the second test stage using a total of 40 accelerometers. A total of seven accelerometers were installed at locations on the bridge spans that were common to both test stages in order to permit the measurements from the two test stages to be combined during post-processing. The sensors and data acquisition system used for the ambient vibration testing included: (1) uniaxial seismic accelerometers (Model 393C from PCB Piezotronics, Inc.), which have a nominal sensitivity of 1 volt/g, a peak measurement range of 2.5 g, a frequency range of 0.025 to 800 Hz, and a broadband resolution of 0.0001 g; (2) a Hewlett Packard Model 8401A VXI mainframe with Model 1432A input modules; (3) Model 481 signal conditioners from PCB Piezotronics Inc.; and (4) a laptop computer. The acceleration measurements were sampled in different test runs at rates that ranged from 20 Hz to 800 Hz, but the majority of data was sampled at 200 Hz for intervals of 900 seconds. The different sample rates were used to study the effect of the recording bandwidth on the

STRUCTURAL IDENTIFICATION OF CONSTRUCTED SYSTEMS 171

identified frequencies. The field measurements were recorded daily for a period of about one month.

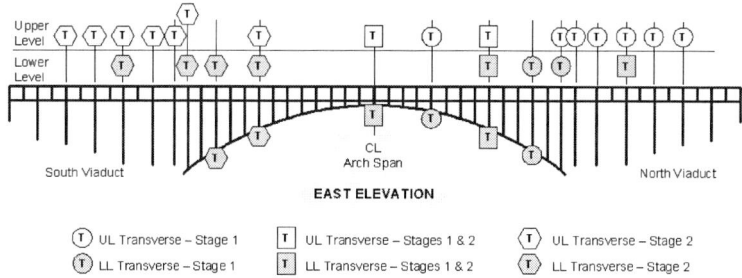

Figure B-3. Accelerometer locations for measuring transverse (lateral) vibrations

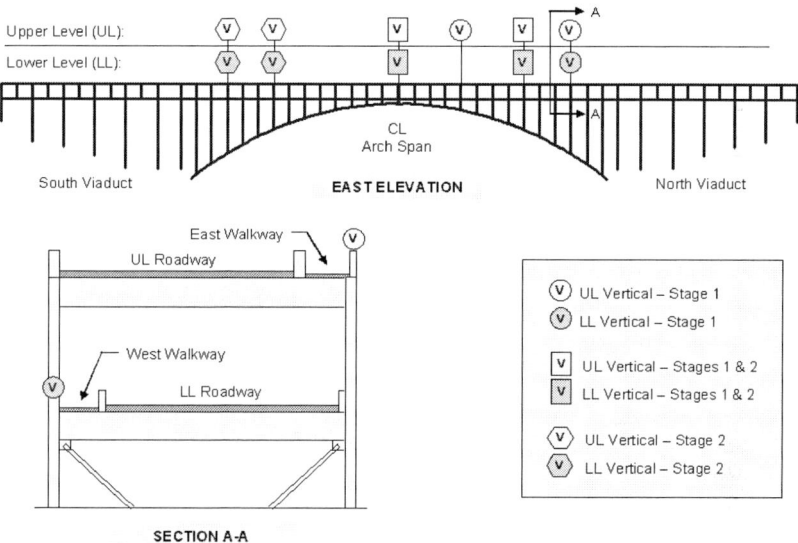

Figure B-4. Accelerometer locations for measuring vertical and torsional vibrations

B.1.5 Data Analysis (Step 4) and Model Calibration (Step 5)

This step aimed to estimate the modal parameters from the measured acceleration responses using output-only modal analysis methods. The data analysis procedure applied to each data file included: (1) visual inspection of time windows from each data set, channel by channel, to identify and exclude channels that were noisy or exhibited blatant measurement errors; (2) application of digital filters to remove any DC bias or drift, and to minimize the influence of high frequency components outside the frequency band of interest; (3) manual removal of spurious noise spikes based on visual inspection of the data records; (4) clustering of time-domain data according to transverse, vertical, and torsional responses; (5) data averaging prior to the generation of pseudo FRFs; and (6) application of multiple parameter estimation algorithms such as the complex mode indication function (CMIF) and poly-reference time domain (PTD) methods.

The test-analysis comparison (Table B-1) indicated that, while certain aspects of the initial FE model appear reasonable (e.g. distribution of vertical stiffness and mass), the model was unable to properly simulate the bridge due to errors related to the absolute mass and stiffness of the main span and lateral continuity conditions between the viaducts and the main span. Given the complexity of the structure and the relatively low spatial resolution of the experimental modal parameters, it is not possible to perform a detailed, formal calibration. Rather, it is more appropriate to use the experimental results together with heuristics and a detailed examination of the structural details in the regions of interest, to identify and modify modeling assumptions to better align the FE model with the experiment.

The first step of the global calibration aimed to increase the vertical bending stiffness of the structure to better align the model with the frequencies identified from the experiment. In the initial FE model, these vertical members were connected to the upper/lower decks and arch rib with "Equal Constraints" that constrained the three translational degrees of freedom, essentially allowing them to act as pinned-pinned members. Examination of these regions however, revealed near-rigid connections with stiffener plates added to enhance moment continuity. To reflect this, all of the vertical members were connected through "Body Constraints" in all six degrees of freedom. In addition, to recognize the increased stiffness of the member in the vicinity of the connection, rigid offsets were added. These adjustments led to an increase in all of the modal frequencies (especially those associated with vertical modes) and the sequence of modes became consistent with the experiment (Table B-1).

Table B-1. Comparison of the first three natural frequencies

Mode	Exp. (Hz)	Initial (Hz)	Diff. (%)	Rev. (Hz)	Diff. (%)	Description
1	0.616	0.512	-16.883	0.588	-4.545	1st lateral bending (arch)
2	0.739	0.505	-31.664	0.721	-2.436	2nd vertical bending
3	0.952	0.890	-6.513	0.973	2.206	3rd vertical bending
4	1.182	0.977	-17.343	1.054	-10.829	2nd lateral bending (arch)
5	1.506	1.257	-16.534	1.404	-6.773	1st lateral bending (global)
6	1.587	1.535	-3.277	1.566	-1.323	1st vertical bending
7	1.732	1.651	-4.677	1.714	-1.039	4th vertical bending
8	2.556	2.393	-6.377	2.505	-1.995	5th vertical bending
9	3.300	3.137	-4.939	3.276	-0.727	6th vertical bending
10	4.110	3.955	-3.771	4.061	-1.192	7th vertical bending

STRUCTURAL IDENTIFICATION OF CONSTRUCTED SYSTEMS   173

The second step in the global calibration was to update the continuity conditions at the expansion joints between the viaduct spans, towers, and main span to better reflect the observed lateral modes. In the initial FE model, the discontinuity at the lower deck was represented by two separate floor beams and the joints along the fascia girders were constrained by "Equal Constraints" in the vertical direction only. Rigid links as well as "Equal Constraints" in both the lateral and vertical direction were used to simulate the tower-deck interface at the upper deck level. Consequently, the movement of the two towers in the longitudinal direction was unconstrained and the initial model displayed pure tower modes just after the first two vertical and lateral modes of the bridge. These modes are inconsistent with the experimental results, which did not indicate any peaks in the mid-height response of the towers below 4 Hz. To better align the initial FE model, the movements in the longitudinal direction were constrained by defining additional "Body Constraints" at both the lower and upper deck levels. Furthermore, the shear and torsion releases in the rigid links at the tower and upper interface were also removed.

After the FE model is updated, its frequencies and mode shapes were calculated and compared with experiment results as shown in (Table B-1) and Figure B-5. It is seen that the differences between frequencies of the initial FE model and those from experiment have been greatly reduced, and mode shapes especially the third lateral one has also been improved to match experiment results through the model updating procedure. By taking advantage of an element-level 3D FE model and ambient vibration testing, this application demonstrated both the potential and limitations of St-Id to provide an in-depth understanding of the physical behaviors of large-scale constructed systems. More importantly, this study illustrated the significant modeling (epistemic) uncertainties that can challenge the reliability of FE models for large constructed systems. Without performing a St-Id of the test bridge, the seismic evaluation and possible retrofit designs would have been underpinned by an FE model that was poorly correlated with the actual response of the structure.

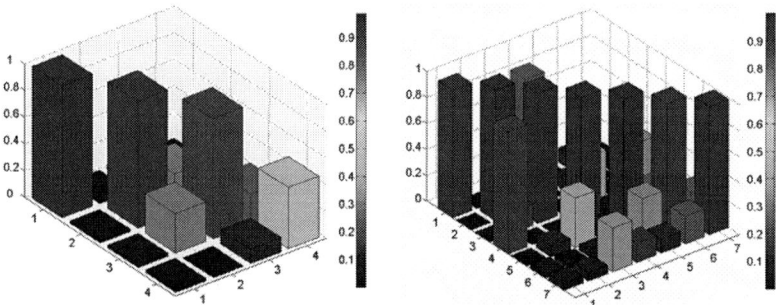

Figure B-5. MAC of experimental and updated analytical (a) lateral modes and (b) vertical modes

B.2 Throgs Neck Bridge

B.2.1 Bridge Description

The Throgs Neck Bridge (Figure B-6) is a major suspension bridge designed by Othmar Ammann across the East River in New York City. This bridge was commissioned on January 1961 and carries approximately 105,000 vehicles per day. The main span of the bridge is 549 m long, with an anchorage to anchorage total length of 886 m. The suspension bridge was designed with an 8.5-meter-deep stiffening truss under the deck, allowing wind to flow through the bridge. Six lanes of vehicular traffic rest on a series of laterally arranged transverse floor trusses, and these transverse trusses are framed into two longitudinal stiffening trusses within the same planes of main cables. Two main cables (976 m in length) support the main span and two side spans.

Figure B-6. The Throgs Neck Bridge

B.2.2 Objective of Structural Identification Application (Step 1)

Structural safety inspection and rehabilitation have been performed several times on the bridge since 1980, which included replacing the roadway decks, repairing the structural steel, modifying expansion joints, replacing existing rocker bearings, improving the drainage system, re-wrapping the main cables, and rehabilitating the electrical systems (Prader et al. 2010; J. Zhang et al. 2009a; J. Zhang et al. 2009b). As part of a seismic vulnerability assessment, the DI3 was charged with performing an ambient vibration monitoring study of the suspended spans and towers of the bridge (Step 1 of the St-Id process).

B.2.3 System Identification Method Development (Step 2)

The three-dimensional finite element model (Figure B-7) for the suspended bridge was constructed by using the commercial finite element code ADINA. The main cables of the bridge are 24-inch diameter. They were modeled using elastic cable elements in ADINA, which only transmitted longitudinal forces and could be used with large displacements. Suspender connecting the main cables and the superstructure was modeled with cable elements as well. All members of the stiffening and floor trusses, lower lateral bracing, and upper lateral bracing were modeled using three-dimensional elastic beam elements with six degrees of freedom at each node. The deck system is a 5-inch concrete filled steel grid, which is supported by W18 x 55 stringers. The concrete filled steel grid was modeled with shell elements having the concrete's elastic modulus, and the stringers were modeled by three-dimensional elastic beam elements. The rigid joint assumption was made in the joint region where struts meet the legs at the top of the tower. The cable saddles are transversely located in an eccentric position with respect to the center of each leg. This was represented by adding a set of nodes at the top of the upper struts with two eccentric nodes for the cable saddle and applying a rigid link constraint to these nodes so that they acted together as a rigid body. The structural mass was determined from the bridge's original construction drawing plans and distributed to each node. The structural model consisted of 15,175 nodes, 294 cable elements for the cables and the suspenders, 9,386 beam elements for the stiffening and floor trusses, the bottom and upper laterals, and the stringers, and 6,048 shell elements for the concrete filled steel grid.

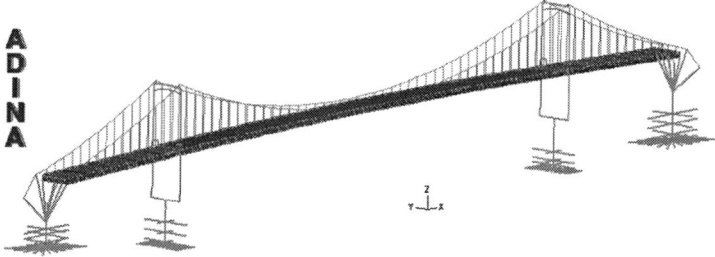

Figure B-7. Finite element model of Throgs Neck Bridge

Careful modeling of structural connections among complex sub-systems is an effective way to reduce uncertainty from the analytical aspect. For instance, the wind tongues at the anchorage are vertical members at the center of the floor truss that bear in the transverse direction into a steel bearing, which is rigidly attached to the anchorage. The wind tongue was represented by 3-dimensional beam elements, rigid links, and gap elements. Each gap element had two nodes, one was slaved to the wind tongue, and the other one was slaved to the anchorage. The two gap elements at both sides of the tongue transmitted any transverse force from the wind tongue directly to the anchorage. Connections at tower-deck and deck-stringer interfaces were also carefully modeled using links and gap elements.

176 STRUCTURAL IDENTIFICATION OF CONSTRUCTED SYSTEMS

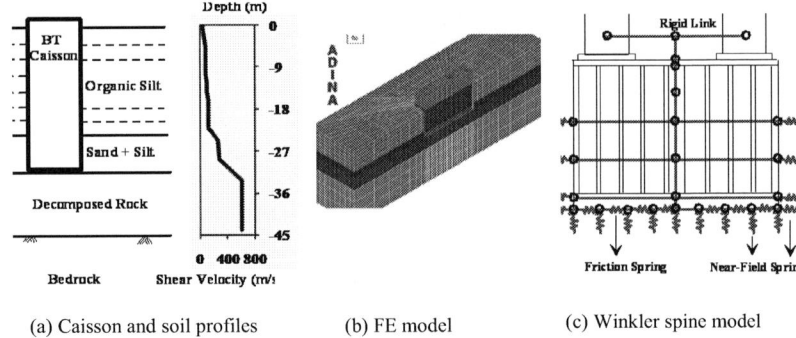

(a) Caisson and soil profiles   (b) FE model   (c) Winkler spine model

Figure B-8. Modeling of the soil-caisson interaction

Special attentions were also put on FE modeling of soil-structural interactions by combining the geotechnical and structural analysis efforts. Bridge towers and anchorages are founded on large rigid reinforced concrete caissons. Unlike pile foundations supporting approach structures of this bridge, caissons are deep foundations with significant weight, and their seismic response will be driven mainly by their inertial interaction with the surrounding soil. Gapping and separation between the caisson and the surrounding soil is a major parameter in the evaluation of the caisson's behavior under seismic loads. For these reasons, the most practical Winkler approach was used to represent the caisson-soil interaction (SSI), and the Winkler spine models were developed to simulate tower and anchorages foundations. According to the Winkler method, a discretization scheme for the caisson-soil system of the global model was first established. Figure B-8a illustrates Bronx Tower (BT) caisson and soil profiles with the soil parameters established from soil exploration. Figure B-8b shows the discretization schemes for Bronx Tower caisson-soil system. As shown, the soil-caisson system in the global model was represented by a combination of three-dimensional elastic beam elements representing the spine of the caisson, constraints (rigid links), and interface near-field SSI elements considering soil plastic property and gapping effect. The non-linear force-displacement property for each near-field SSI element was calibrated by comparing the nonlinear pushover analysis results of the Winkler spine model (Figure B-8c) with those from the FE simulation of the soil-caisson system (Figure B-8b). Detailed FE analysis of the soil-caisson system considered both the plastic behavior of the soil and also simulated the caisson-soil gapping behavior, i.e., geometric non-linearity.

Therefore, the calibrated Winkler spine mode was able to accurately capture the nonlinear behavior of caisson-soil interaction. A similar discretization scheme was also established for Bronx Anchorage, Queens Anchorage and Queens Tower. Eigenvalue analysis of the constructed 3D FE model includes the near-field SSI elements was done with the aim of identifying the fundamental periods and mode shapes of the structure, and its results were compared to those from the ambient vibration tests on the bridge which would be described later.

B.2.4 Experimental Studies (Step 3)

A roving instrumentation scheme was utilized in the ambient vibration tests of the Throgs Neck Bridge. It is employed instead of an instrumentation scheme in which all accelerometers remain at stationary locations on the structure throughout the testing program for the following reasons: (1) limited numbers of available sensors or data acquisition channels, (2) the capability to rapidly characterize the structure at many measurement stations, and (3) the ability to estimate mode shapes with a relatively fine spatial resolution. The instrumentation scheme developed for the test bridge incorporated multiple stationary reference locations with a number of roving accelerometers (Figure B-9). A total of forty five unidirectional accelerometers were used in this scheme to measure the vibration responses. Of this total, thirty accelerometers remained at stationary positions throughout the testing, and served as reference sensors. The remaining fifteen accelerometers were roved across the spans and towers of the bridge to measure the structural responses in these regions with added spatial resolution in four roving setups. This instrumentation scheme enabled multiple and spatially distributed reference sensors to be utilized with the vibration data analysis performed for each region of the bridge (side spans, main span, and towers). It was expected that having multiple and spatially distributed reference locations available for the data analysis and interpretation stages would enable a significantly more stable and reliable identification of the dynamic characteristics than would be possible from a conventional roving setup with only a few reference locations.

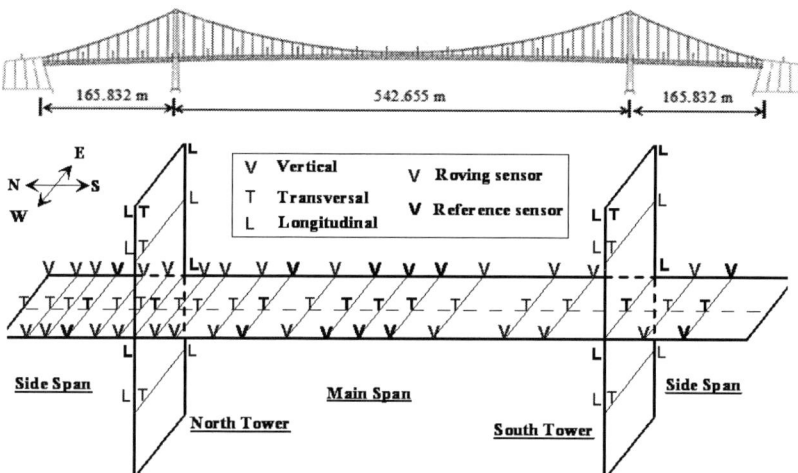

Figure B-9. Accelerometer layout on bridge deck and tower

178 STRUCTURAL IDENTIFICATION OF CONSTRUCTED SYSTEMS

The field testing stage of the vibration testing program included both the installation and verification of the operation of the accelerometers, cabling, and data acquisition components. The accelerometers were installed on the steel bridge members at their designated locations using magnetic sensor mounts. The individual cables for each accelerometer were installed and temporarily secured along the bottom chord of the stiffening trusses.

Careful execution of the instrument installation and on-site quality control may reduce the uncertainty by identifying the possible causes of error and their impact on the measurements. For instance, calibrating each sensor individually and the sensory system as a whole in the field environment may reduce the systematic and aleatory errors in the experiment stage. After all the accelerometers and cables were verified, the data acquisition system was set to record one hour long data sets throughout the duration of the ambient vibration test program. The vibration measurements were sampled at 200 Hz which was at least 20 times greater than the estimated maximum frequency of interest.

An automated data pre-processing procedure was first developed and implemented to reduce bias and aleatory errors affecting the data quality (Figure B-10). It consists of visual inspection, time window selection, digital filtering, cross-correlation construction, windowing, and data averaging in time or frequency domain. After test data were cleaned, three data processing methods, including the Peak Picking, PolyMax, and CMIF methods, were performed independently for modal parameter identification. Correlation analysis of the identified results from these methods was used to verify the reliability of the identified results and provide the bridge owner more confidence in using the identified results for decision making. Different test data sets were investigated independently by various post-processing methods to make sure that the identified results are reliable. Multiple reference measurements also provided an effective means to identify the most likely candidates for the natural frequencies and mode shapes by visually comparing the consistency of the identified shapes constructed from different reference locations.

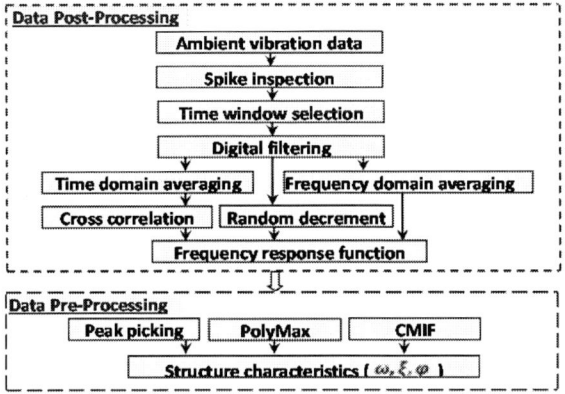

Figure B-10. Flowchart of data processing procedure

STRUCTURAL IDENTIFICATION OF CONSTRUCTED SYSTEMS 179

It is important to note that the dynamic characteristics of any suspension bridge structure should be expected to vary by seasonal and environmental conditions. It is desirable to repeat data collection over different wind and environmental conditions, especially during the extremes of ambient temperature and solar radiation conditions in order to understand the variability of dynamic properties.

B.2.5 Data Analysis (Step 4) and Model Calibration (Step 5)

Measurements from the pseudo-roving of instruments installed in sub-regions of the bridge (north side span, south side span, main span, north tower, and south tower), together with the measurements from the large number of reference channels, were used to perform the span structural characterization. The identified frequencies and modes shapes were compared with those from the eigenvalue analysis of the 3D FE model. Structural frequencies identified from independent post-processing methods (the Peak Picking, PolyMax, and CMIF methods), as well as the FE analysis, are compared in Table B-2, which shows favorable correlation for a wide range of natural periods from 0.50 seconds to 6.5 seconds. For illustration purpose, the first six vertical mode shapes of the main span and the side spans identified from the PolyMax methods are illustrated in Figure B-11, which are comparable with the mode shapes from the FE analysis. Mode shapes in the transverse directions from ambient tests and FE analysis are also comparable, though not provided here for conciseness.

The seismic design criteria required that the first three vertical modes and first three horizontal modes shall be compared to those obtained from ambient vibration measurements of the bridge for verification of the structural model. The constructed FE model apparently satisfies the requirements of the Seismic Design Criteria and is ready to be used as a tool for bridge maintenance related decision-making.

Table B-2. Analysis of lateral and vertical mode frequencies

Mode	Experimental (Hz)			Finite Element (Hz)
	PolyMax	PP	CMIF	
1st vertical	0.187	0.183	0.195	0.189
2nd vertical	0.216	0.219	0.219	0.218
3rd vertical	0.310	0.305	0.305	0.310
4th vertical	0.391	0.378	0.378	0.444
5th vertical	0.444	0.440	0.439	0.443
6th vertical	0.624	0.623	0.610	0.610
1st transverse	0.150	0.159	0.146	0.152
2nd transverse	0.467	0.476	0.512	0.483
3rd transverse	---	0.794	0.793	0.756
4th transverse	1.009	1.025	1.025	1.054
5th transverse	1.451	1.538	1.550	1.630

180   STRUCTURAL IDENTIFICATION OF CONSTRUCTED SYSTEMS

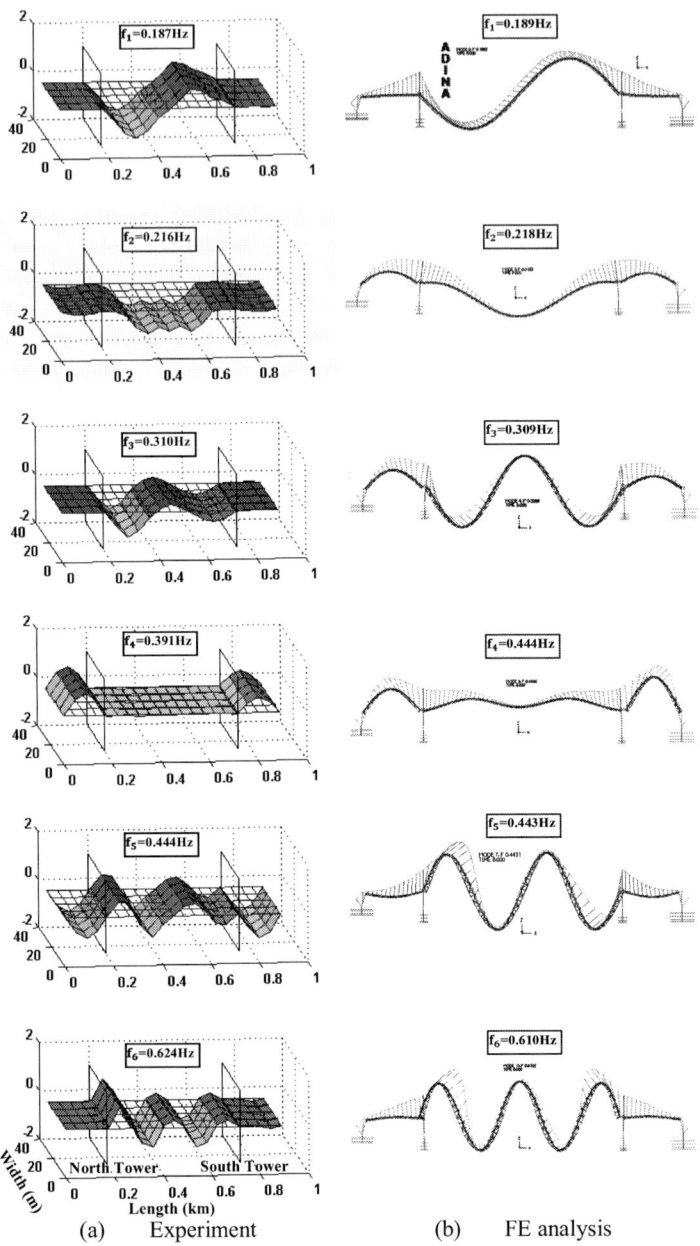

(a)   Experiment   (b)   FE analysis

Figure B-11. Mode shape correlation

STRUCTURAL IDENTIFICATION OF CONSTRUCTED SYSTEMS 181

B.3 Golden Gate Bridge

B.3.1 Bridge Description

The Golden Gate Bridge at the entrance of the San Francisco Bay was completed and opened to traffic in 1937. The bridge has a 1280 m (4200 ft) long main-span and 343 m (1125 ft) side-spans. Two stiffening trusses support an orthotropic roadway deck and horizontal planes of wind bracing system at the top and bottom plane of the truss chords. The legs of the towers, 210 m (745 ft) above the water level, have cellular box sections, connected by horizontal struts at seven elevations (Stahl et al. 2007). Figure B-12 is a view of the north tower and the roadway from the second strut on the south tower.

As a distinguished long-span bridge, the bridge has been the subject of several instrumentation studies in the past. The U.S Coast and Geodetic Survey installed seismological sensors on the piers, towers, deck and cables between 1933 and 1942 at different stages of construction and initial operation (Vincent 1958). Ten vertical accelerometers were installed in 1945 and operated until 1954 (Vincent 1962). Abdel-Ghaffar and Scanlan (1985) performed the most recent ambient vibration study using accelerometers on the main-span and south tower, and characterized the response of the bridge to ambient wind, wave and traffic excitation via estimation of mode shapes, and the associated natural frequencies and modal damping ratios.

Figure B-12. A view of the Golden Gate Bridge from the North Tower

182 STRUCTURAL IDENTIFICATION OF CONSTRUCTED SYSTEMS

B.3.2   Objectives of St-Id Application

The objective of this study is to present a statistical analysis of the vibration modes of the Golden Gate Bridge using ambient acceleration data obtained from large-scale deployment of a wireless sensor network (WSN). The contribution is to demonstrate that the spatial and temporal sensing possible with WSNs provides high resolution and confidence in the identified vibration modes.

In contrast with the earlier studies, the WSN in the present study has a much higher spatial resolution and the ambient vibration data is collected over an extended period of time, which allows statistical analysis. The statistical approach demonstrated in this study is applicable to sensor networks on other bridges and buildings and can be used as a template for analysis of systems with streaming data. The ambient vibration data are used to:

1. Identify the modal properties of the bridge including the higher modes.
2. Establish the confidence intervals for the estimated parameters.
3. Compare the modes from a finite element model of the bridge with the confidence intervals.
4. Compare the estimated parameters from a previous deployment of the sensors on the bridge with the confidence intervals.
5. Compare the estimated parameters using spectral methods, which are less expensive alternative modal identification algorithms, with the confidence intervals.

B.3.3   Experimental Program

A wireless sensor network was designed, developed, and deployed to measure and record ambient accelerations of the bridges. The network was designed to be scalable in terms of the number of the nodes, complexity of the network topology, data quality and quantity by addressing integrated hardware and software systems such as sensitivity and range of micro-electro-mechanical-systems (MEMS) sensors, communication bandwidth of the low-power radios, reliability of command dissemination and data transfer, management of large volume of data and high-frequency sampling (Pakzad et al. 2008). Each node has a sensor board with MEMS accelerometers in two orthogonal directions, a temperature sensor, and a micro-controller and communication mote. The nodes on the main-span measure acceleration in vertical and transverse directions. On the tower, the nodes measure acceleration in transverse and longitudinal directions. To study the cost-performance tradeoffs for MEMS accelerometers, each node has two sensors in each direction. The ADXL202 (Analog Devices 2011) accelerometers have a range of ± 2 g. For low-level motion, the Silicon Design 1221L (Silicon Designs 2007) is used with a range of ± 150 mg. The wireless sensor network is controlled by a high-level program for the TinyOS software platform (TinyOS 2010). The software architecture and extensions to TinyOS are described in Kim et al. (2007).

For the Golden Gate Bridge, the wireless sensor network consisted of 64 nodes on the main-span and the south tower, for a total of 320 sensors, including

STRUCTURAL IDENTIFICATION OF CONSTRUCTED SYSTEMS   183

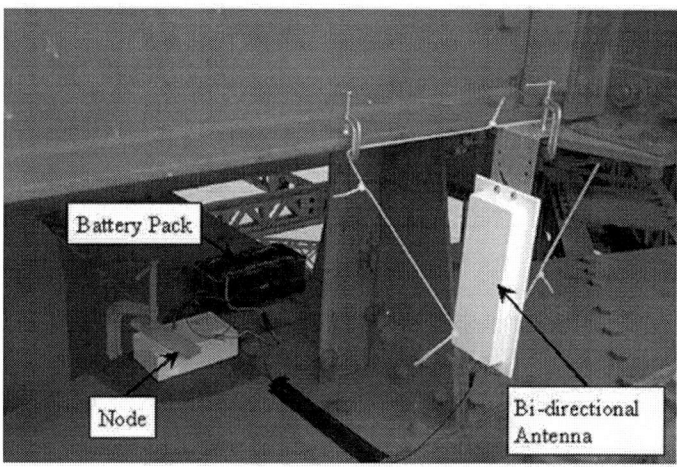

Figure B-13. Node with its battery pack and bi-directional antenna on the main-span of the Golden Gate Bridge

high-level accelerometers, low-level accelerometers, and thermometers. Fifty-six (56) of these nodes were installed on the main-span of the bridge, which provide a network that is capable of identifying up to 50 modes in each sensing direction. Considering the low-energy of the ambient vibration for higher modes, the density of the network is sufficiently high for an expanded modal analysis of the bridge.

Figure B-13 shows a node package including its battery and bi-directional antenna, installed on the main-span of the bridge. Figure B-14 shows the instrumentation plan for the main-span. The nodes on the 1280 m (4200 ft) main-span were located at 30.5 m (100 ft) spacing, but a 15.25 m (50 ft) spacing was used where an obstruction hindering radio communication. The eight nodes on the 210 m (745 ft) south tower were placed at the ends of four struts above the roadway. Fifty-three (53) nodes were installed beginning on July 10, 2006, on the west side of the main-span. On September 15, 2006, the batteries were replaced for the nodes on the main-span and three extra nodes were added on the east side. There were a total of 174 data collection runs of the network during the deployment, which lasted until October 14, 2006, including testing and debugging.

The sampling rate for all runs was 1 kHz, but since the significant vibration frequencies of the bridge are much lower, the data were averaged on the node and downsampled to 50 Hz prior to transmission. The averaging is very effective in reducing the noise level and improving the accuracy of the estimated parameters. In some of the runs all five channels on a node (two high-level motion sensors, two low-level motion sensors and the temperature sensor) were sampled, but in other runs the channels were limited to the low-level accelerometers to reduce the volume of data. Each run generated up to 500 kB data per node, which for the network of 60 nodes produced 30 MB data for 15 million samples. Approximately 1.3 GB data was collected during the deployment of the wireless sensor network on the Golden Gate Bridge.

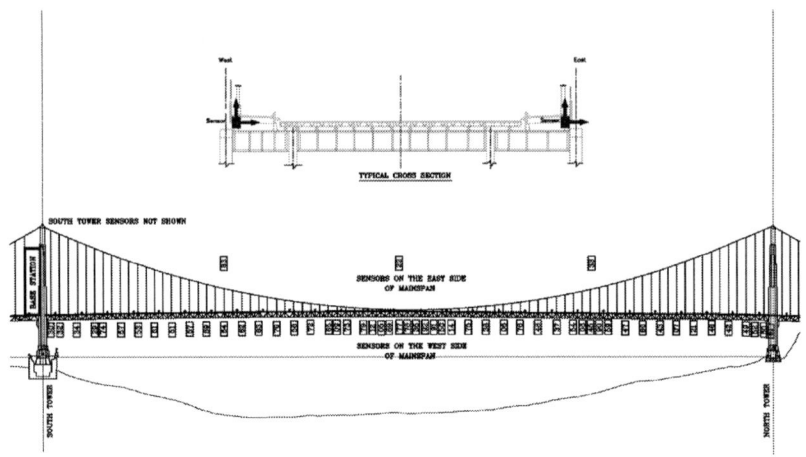

Figure B-14. Instrumentation plan for 56 acceleration sensor nodes on the main-span of the Golden Gate Bridge

B.3.4   System Identification and Data Analysis

Collecting ambient vibration data from many runs makes it possible to compute statistical measures for the identified vibration modes. In this section a statistical analysis of the vibration modes of the Golden Gate Bridge is presented. The vibration frequencies, damping ratios and mode shapes for the main-span are identified using the ARMA method and stabilization graphs for data from all 174 runs of the network. Then this information from multiple runs is used to make statistical inference about the modes and to obtain probabilistic estimates of the modal properties. The confidence intervals are compared with the results from other methods to make statistical inference about the accuracy of the identified vibration modes.

The statistical analysis includes histograms and confidence intervals of vibration frequencies, damping ratios, and mode shapes. In each case the statistics are also presented by the mean value of the parameters and their confidence intervals (CI). A 95% CI for a point-estimated parameter can be interpreted as an interval that is believed, with 95% confidence, to include the true value of the parameter. In other words, if the same procedures are repeated (sampling from the population, estimating the parameter and finding CIs), 95% of the times the estimated CIs are expected to include the true value of the parameter. This statistical analysis allows inference about the certainty of the estimation of modal vibration properties. The narrower the histograms or confidence intervals are, the less uncertainty the estimated values have.

The approach can also be used as a comparison basis for other estimations of the same modal parameters. A new estimate that lies inside the CI is consistent with the hypothesis that no change has occurred.

The results of system identification are used to estimate statistical properties of vibration frequencies, damping ratios and mode shapes for the vertical, torsional and transverse modes of the main-span. The histograms for vibration frequencies and damping ratios are plotted with the mean value centered at the origin over a range of ±3.5-times the standard deviation. Figure B-15 shows a sample of the histograms of the vibration frequencies and the damping ratios for the vertical modes. Overall, twenty-five (25) vertical, nineteen (19) torsional and twenty-three (23) transverse modes with frequencies less than 5 Hz are identified. The vertical axes of the graphs are the identified mode numbers. The mean and standard deviation of the estimated parameters are listed for each mode. Each histogram is marked with an "A" for the anti-symmetric modes or an "S" for the symmetric ones.

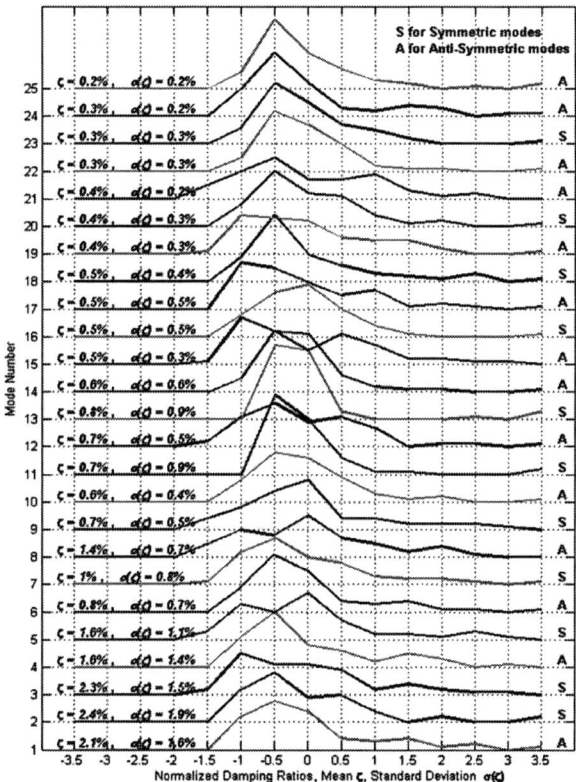

B-15. Histograms of identified vertical damping ratios of the main-span

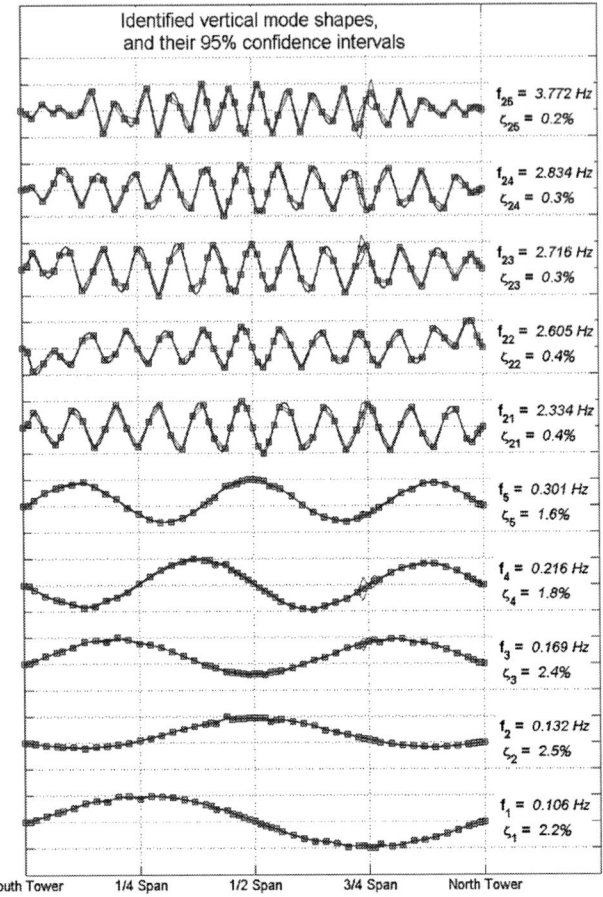

B-16. Identified five lowest and highest frequency vertical mode shapes of main-span and their 95% confidence intervals

For mode shapes, a sample of the mean value and 95% confidence intervals of each node for the lowest five vertical modes are plotted in Figure B-16. The confidence intervals are small for both lower and higher mode shapes, which is an indication of the high quality of data within the frequency range up to 5 Hz.

B.3.5 Comparison and Interpretation

As examples of using the confidence intervals, three sets of identified vibration modes of the Golden Gate Bridge are compared with the statistical results. These data sets were obtained using different identification procedures with data from earlier deployments of accelerometers, finite element models, and the peak picking methods.

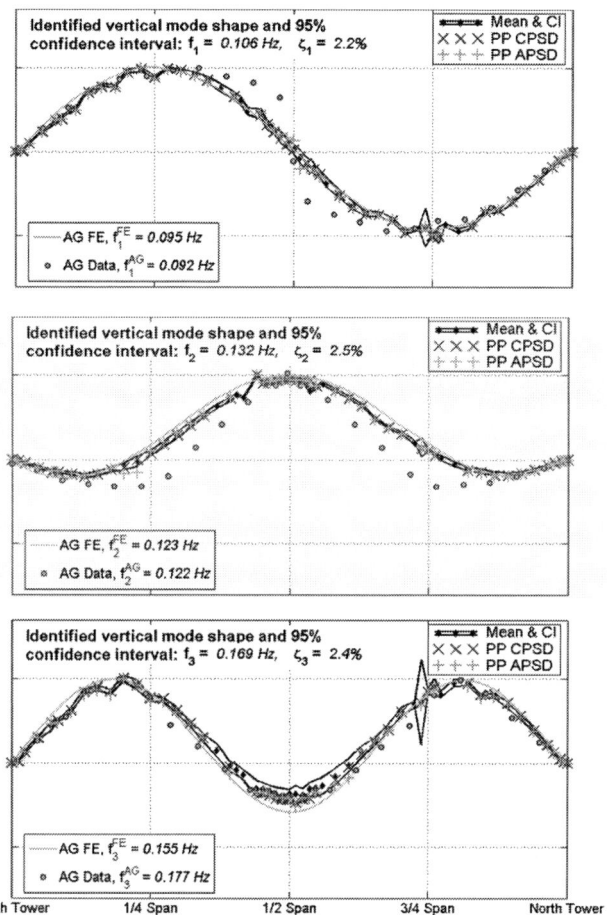

Figure B-17. Comparison between the lowest three identified vertical mode shape using ARMA method, Peak Picking method, 1982 data from Abdel-Ghaffar test, and 1982 finite element model by Abdel-Ghaffar

Abdel-Ghaffar et al. (1985) used sensors on the south side of the main-span and the south tower of the Golden Gate Bridge. On the main-span, six accelerometers, two in each direction, were installed at a location near the south ¼-span as the reference station. Another group of six accelerometers were moved from station to station at 18 locations and the spectral amplitude of each was compared with that of

188   STRUCTURAL IDENTIFICATION OF CONSTRUCTED SYSTEMS

the reference station to identify the vibration modes of the bridge. This is a peak-picking method, in that the relative spectral amplitudes at peak frequencies are used to estimate the mode shapes, but since the data were not collected simultaneously it can produce large errors. Two-dimensional finite element models of the bridge were also used to compute vibration modes (Abdel-Ghaffar and Rubin 1983a; Abdel-Ghaffar and Rubin 1983b). Two variations of the peak picking methods, one using the auto power spectral densities, and the other using cross power spectral densities are also used to identify the vibration modes of the data from one run (#174) of the current wireless sensor network.

Figure B-17 shows the lowest three identified vertical modes and their confidence intervals. All of the three presented mode shapes from peak picking methods lie within the confidence intervals, which confirms the accuracy of the methods. These results suggest that for the goal of estimating modal properties of a long-span bridge structure, the simpler and faster peak picking methods are sufficient.

The identified vibration modes (Abdel-Ghaffar and Scanlan 1985) are generally similar to the statistical analysis of the data from the wireless sensor network. Some of the mode shape ordinates, however, are outside the confidence intervals. In this comparison, it must be recognized that the bridge has been retrofitted since the earlier data was collected including a complete replacement of the roadway deck. This comparison indicates that the earlier estimates of mode shapes are not very accurate because of the change in the main-span, but also because of the quality of the data and the error in the system identification method for the earlier estimates. The 1985 data best matches the confidence intervals for higher frequency vertical modes (not shown here), indicating that the change in dynamic properties of the bridge most affected the low frequency modes.

In the case of FE model (Abdel-Ghaffar et al. 1985), all the modes except for the second transverse mode (not shown) fall within the confidence intervals. The confidence intervals for second transverse mode are wider than the other modes, and have a larger spatial variation, suggesting that the collected data is noisier at this frequency, which has resulted in confidence intervals that are less accurate.

B.4   Vincent Thomas Bridge

B.4.1   Bridge Description

The Vincent Thomas Bridge (VTB) is located in the metropolitan Los Angeles, California. The bridge was constructed in 1963, connecting two major harbors in the US, the Port of Los Angeles and the Port of Long Beach. The VTB is a cable-suspension bridge, approximately 1850 m long, consisting of a main span of 457 m, two suspended side spans of 154 m each, and two ten-span cast-in-place concrete approaches of 545 m length on both ends. The roadway is 16 m wide and accommodates four lanes of traffic. A major seismic retrofit was performed between 1996 and 2000, including a variety of strengthening measures, and the incorporation of about forty-eight large-scale nonlinear passive viscous dampers. A photo of the VTB is shown in Figure B-18.

STRUCTURAL IDENTIFICATION OF CONSTRUCTED SYSTEMS   189

A web-based, real-time, continuous monitoring system has been developed to monitor the VTB by the researchers at the University of Southern California (USC). Using twenty-six (26) high performance accelerometers owned by the California Geological Survey (CGS), the developed bridge monitoring system is based on a multi-thread software design. The software consists of three main threads: (1) data acquisition thread (publisher), (2) data transceiver thread (server), and (3) local monitoring thread (clients). The highly efficient software architecture enables the monitoring system to acquire raw sensor data with multi-channels from the bridge site, to process the data at the server in USC, and to distribute the data to multiple clients simultaneously at different locations over the Internet. The sensor locations on the bridge and a schematic of the system architecture of the developed bridge monitoring system are shown in Figure B-19. A detailed description of the web-based bridge monitoring system can be found in several studies (Masri et al. 2004; Wahbeh et al. 2005; Yun et al. 2007).

Figure B-18. The Vincent Thomas Bridge (Source: Yun et al, 2007; reproduced with permission from John Wiley and Sons)

190   STRUCTURAL IDENTIFICATION OF CONSTRUCTED SYSTEMS

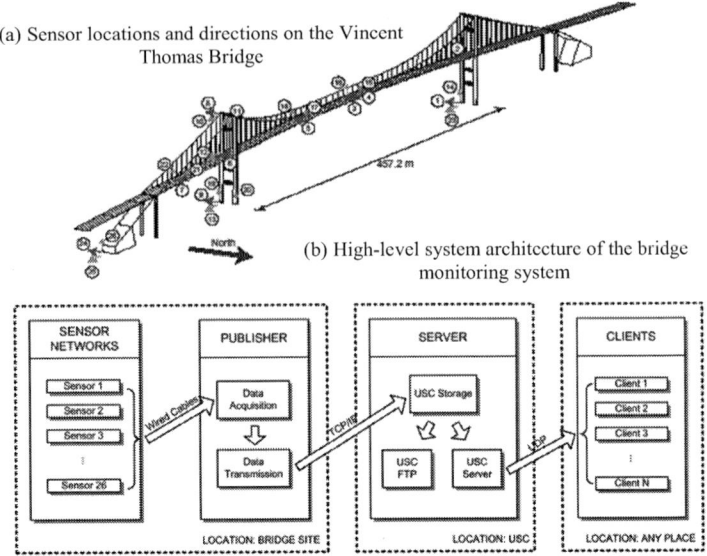

Figure B-19. USC web-based, real-time, continuous monitoring system for the Vincent Thomas Bridge (Source: Yun et al, 2007; reproduced with permission from John Wiley and Sons)

B.4.2   Objective of St-Id Application

As described earlier, the VTB has been monitored with twenty-six (26) accelerometers. These sensors were installed by the California Strong Motion Instrumentation Program. The CSIMP has been in operation since 1972, and the purpose of this program is to record the strong shaking of the ground and in structures during earthquakes for the engineering and scientific communities through a statewide network of strong motion instruments. As a consequence of the program, many important civil structures in California are provided with seismic sensors connected to ruggedized data loggers. The recording of the sensor data, however, is *trigger-based*, so that the raw sensor data collected in the temporary memory buffer of the data logger are analyzed, and if the raw data are characterized as seismic signals, then the data are saved on permanent data storage devices of the logger. Therefore, valuable data from ambient and abnormal loading conditions other than earthquakes, such as ship-bridge collision, are wasted. Not only seismic data, but these *non-seismic* data are also very important in the analyses of structural performance to protect structures from various hazards.

The reason for the trigger-based recording is obvious: for continuous monitoring, the dynamic sensor measurements, involving a number of sensors at high frequency sampling rates require a large data storage spaces. This problem could be

STRUCTURAL IDENTIFICATION OF CONSTRUCTED SYSTEMS 191

overcome with the web-based, continuous bridge monitoring system described earlier, combined with advanced *on-line* structural identification (St-Id) techniques.

The main objective of the St-Id application for the VTB is to quantify the structural performance of the bridge when subjected to various loading conditions and significant environmental effects, which is infeasible with traditional visual bridge inspection approaches. Employing the continuous monitoring system and on-line St-Id, *physically meaningful information* could be extracted from the complicated, raw time histories of *raw sensor data*.

It should be noted, however, that the traditional visual inspection is not intended to be replaced with the continuous monitoring system and on-line St-Id. Since 1972, visual inspection has played a critical role in the National Bridge Inspection Program (NBIP) and has been considered a major approach to evaluate bridge conditions. Using both the traditional visual inspection approaches and modern technologies of sensing and advanced data processing, the understanding of our valuable bridge structures could be extended. This is especially true for the VTB with a coarse sensor network of twenty-six accelerometers, since with this number of sensors the localized damage can hardly be found due to the lack of spatial resolution in the sensor network.

B.4.3   Bridge Identification Using Seismic Vibration Data

The VTB is located in the world's most active seismic zone of Southern California and numerous small and moderate-size earthquakes occur frequently in this region, as well as less-frequent, destructive, large-size earthquakes. Using the web-based bridge monitoring system, the small and moderate-size earthquakes can be used as excellent bridge shakers, exciting many dynamic modes, which are not excited in ambient vibration conditions.

In the early morning hours of 22 February 2003, an earthquake (magnitude M=5.4) occurred in the vicinity of the city of Big Bear, California. The epicenter was located about 180 km from the VTB, and no bridge damage was reported after the earthquake. Using the 26 accelerometers installed on the VTB, the seismic vibration data were obtained with the bridge monitoring system. Sample time histories of the seismic vibration at the base and deck of the VTB are shown in Figure B-20.

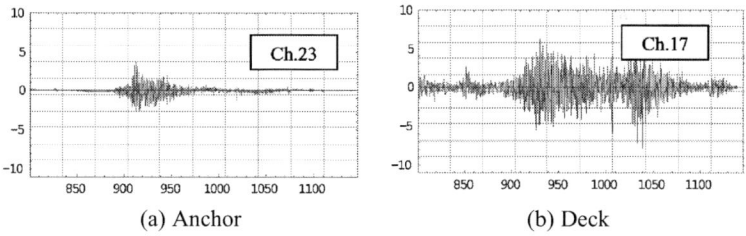

Figure B-20. Sample time histories of the seismic vibration at the base and deck of the Vincent Thomas Bridge

The modal parameters were identified using the LS method, and compared with those for different earthquakes: Whittier (1987) and Northridge (1994). Detailed discussion of this method can be found in Masri et al. (1987). The identified modal parameters agreed for the three earthquakes, which implies that the structural characteristics of the bridge can be reliably identified using the seismic vibration data (Wahbeh et al. 2005).

B.4.4 Forensic Study of Ship-Bridge Collision Accident Using St-Id

On 27 August 2006, the VTB was struck by a large cargo ship, passing under the bridge. The traffic was stopped right after the incident, and the bridge engineers from the California Department of Transportation (Caltrans) investigated possible damage due to the collision. Moderate damage on the maintenance scaffolding at the main span of the bridge was observed through the visual inspection, and the traffic was resumed about 2.5 hours after the collision. This incident, however, left the transportation authorities wondering about the structural integrity of the bridge that is not quantifiable with the visual inspection. Consequently, this study was motivated the need to validate decisions based on the traditional visual inspection approaches with the analysis results based on the more sophisticated St-Id techniques.

A forensic study was performed to assess the structural condition of the bridge before, during and after the collision accident to detect whether significant changes occurred in the bridge vibration signature. Using the acceleration measurements (Figure B-21), the bridge was identified using the Eigensystem Realization Algorithm (ERA), combined with the Natural Excitation Technique (NExT) (Nayeri et al. 2007).

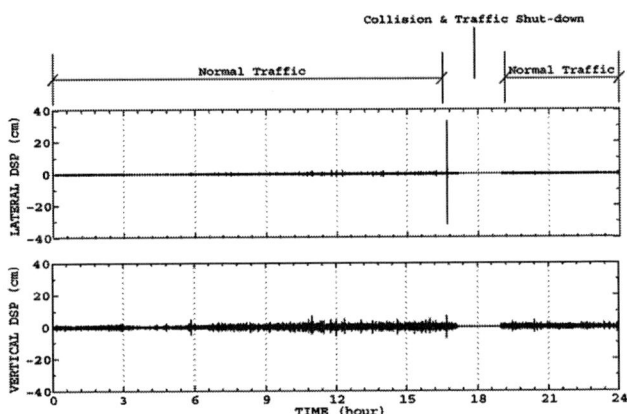

Figure B-21. Twenty-four hour displacement (lateral) time history of the bridge deck on the day of the ship-bridge collision (Source: Yun et al, 2007; reproduced with permission from John Wiley and Sons)

The modal parameters were determined using the NExT-ERA method, and compared for before, during and after the accident. The identified parameters showed

STRUCTURAL IDENTIFICATION OF CONSTRUCTED SYSTEMS 193

that there existed no significant difference between the cases. Sample identified model parameters are illustrated in Figure B-22.

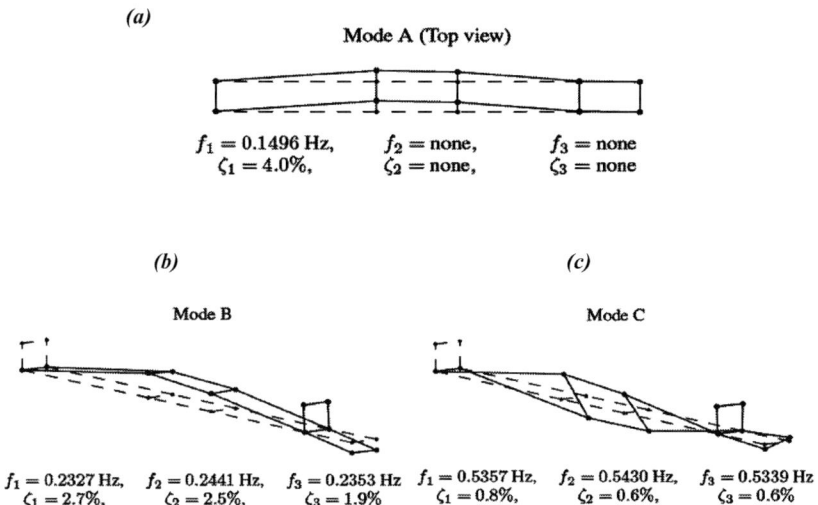

Figure B-22. Sample identified modal parameters of the Vincent Thomas Bridge for the ship-bridge collision accident (Source: Yun et al, 2007; reproduced with permission from John Wiley and Sons)

B.5 Hakucho Suspension Bridge

B.5.1 Bridge Description

The Hakucho Bridge in Japan is a three-span suspension bridge with the total length of 1380 m consisting of 720 m center span and two symmetric side spans of 330 m. Both side spans and the center span are simply supported at the towers. The girder is made of a streamlined steel box girder with a width of 23 m and a maximum web height of 2.5 m. The bridge pylons are made of steel box girder and connected by welding. Both towers are 131 m high and 21 m wide. The construction of the bridge was started in 1985 and ended in 1998. It was finally opened to the public on June 13, 1998. After the construction completion, a series of dynamic tests were performed, including ambient vibration tests.

B.5.2 Objective of Structural Identification Application

Monitoring for wind-induced vibration has been employed especially for long span bridges in Japan. Portable as well as embedded sensor networks are utilized to measure wind condition and dynamic response of the bridge. Such a measurement system was employed at Hakucho Bridge (Figure B-23). The initial objective of the

194 STRUCTURAL IDENTIFICATION OF CONSTRUCTED SYSTEMS

Figure B-23. Hakucho Suspension Bridge

Hakucho Bridge monitoring was to verify the results of wind tunnel test, especially concerning the aerodynamic forces. At the time of the test, bridge engineers in Japan mainly relied on wind tunnel test and there was little effort to confirm the result with a full-scale monitoring of the real bridge. In fact, this was the first initiative in Japan to monitor a long-span bridge with a very dense-array of sensor (i.e., forty measurement points on one half-span of the bridge) for over two weeks.

B.5.3   Instrumentation and Ambient Vibration Monitoring

Densely distributed accelerometers were placed at various locations. On the girder, twenty-one accelerometers were installed with the spacing of 30 m on the main span and of 55 m on the side span near the Jinya approach. Of these 21 accelerometers, 17 were placed on the centerline of the bridge deck, while the other four were mounted on both sides of the middle span to cover the torsional motion of the bridge, as illustrated in Figure B-24. In order to measure wind velocity, an anemometer was installed at the center of the span of the deck. Since the bridge is located at the port entrance, wind orthogonal to the bridge is relatively strong (Siringoringo and Fujino 2008b).

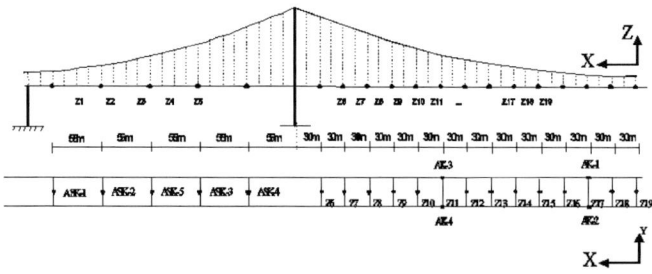

Figure B-24. Sensor layout for wind-induced vibration monitoring of Hakucho Bridge

B.5.4 Data Analysis and System Identification

The measured data clearly show the quadratic relationship between wind velocity and response. The applied structural identification method consists of two steps: identification of vibration modes and inverse analysis of structural properties from the identified modes. For modal identification, the method treats the structure as a multi-input-multi-output system, distinguishing noise from true modes and employing ambient vibration measurement. Two time-domain system identification approaches were applied that is Random Decrement with Ibrahim Time Domain (RD-ITD) and the Natural Excitation Technique with Eigensystem Realization Algorithm (NExT-ERA) (Siringoringo and Fujino 2008b). Initial data analysis shows that identification of higher order modes using RD-ITD is not as straightforward as identification using NExT-ERA. The RD-ITD requires filtering out the low frequency component to obtain higher frequency modes. This procedure is clearly time consuming and not deemed reasonable to be applied to 200 h of measurement data. Therefore, only the NExT-ERA technique is further employed for structural identification.

B.5.5 Structural Identification and Interpretation

Despite variations of natural frequencies and damping ratios, there seem to be clear trends between frequencies, damping ratios and acceleration amplitudes (and consequently, the wind speed). Results show that in general the natural frequencies decrease as the wind velocities increase and damping ratios increase as the wind velocities increase. The variations of natural frequencies and damping ratios are more apparent in the low-order modes as evident by the slopes of the linear trend. The mode shape components reveal two different trends. The real parts of mode shape vectors do not exhibit a distinct trend, indicating no obvious changes. The modal phase angle computed from the imaginary part of the mode shape vectors, however, revealed a clear trend. It was observed that the phase difference is large when the root-mean-square of acceleration (rms) is very small and decreases when the acceleration rms becomes large. These phase differences indicate that the system is nonproportionally damped. The locality effect of phase difference that was concentrated mainly at the edge of girder suggests the contribution of additional damping and stiffness caused by friction force at the bearings. In addition, the decrease and increase of natural frequencies and damping ratios indicated the effect of aerodynamic force along the girder. To study the extent of these effects, they were modeled as additional stiffness and damping: (1) located at the edge of the girder to represent the friction force at the bearings and expansion devices and (2) distributed alongside the girder to illustrate the aerodynamic forces.

For the identification of structural properties an inverse analysis was conducted (Nagayama et al. 2005). Using the inverse analysis contribution of aerodynamic and friction force were with respect to wind velocity were quantified. The results suggest the contribution of aerodynamic forces was much smaller than the effect of the friction force at the bearing. The aerodynamic force contribution is on the order of one-percent when compared to the contribution of the friction force, and its behavior is in agreement with the aerodynamic force obtained from wind tunnel results. Furthermore, the additional damping and stiffness due to the friction force display clear trends: small damping and large stiffness during low-amplitude vibrations. When the wind speed increases the damping also increases, which is when the bearings are unstuck, whereas the stiffness is decreasing as the result of increasing flexibility of the structure (Figure B-25).

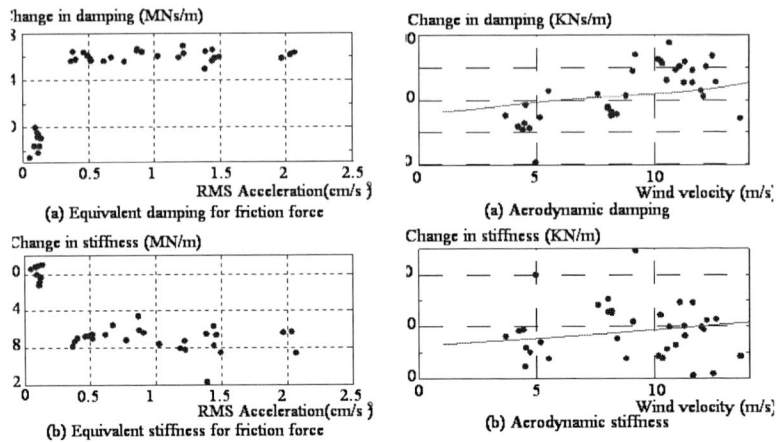

Figure B-25. Identified changes in aerodynamic force and other critical parameters

B.6 Yokohama Bay Bridge

B.6.1 Bridge Description

The Yokohama Bay Bridge (Figure B-26) is located at the entrance of Yokohama Harbor in Japan. It is a continuous three-span cable-stayed bridge with the main girder consisting of a double-deck steel truss-box. The central span is 460m with side spans of 200m each. The upper and lower deck have 6 and 2 lanes respectively, with the upper deck being part of the Yokohama Expressway Bay shore Route and the lower deck a part of the National Route. The bridge has two H-shaped towers of 172m in height and 29.25m in width with a welded monolithic section. Earthquake resistance of the bridge was carefully reviewed in design. Considering the possibility of a large event like the Great Kanto earthquake that Tokyo and Yokohama experienced in 1923, the weak ground and high center of gravity of the bridge, the girder is suspended from the towers and end-piers with link bearing in such a way that the effect of the superstructure on the substructure during an earthquake is reduced. Thus the bridge maintains a long fundamental period of about 7.7 seconds in the longitudinal direction. The structure, with its long natural period, is expected to see lower acceleration during an earthquake but, on the other hand, its displacement is increased. Therefore to restrict the horizontal displacement during earthquake, short

STRUCTURAL IDENTIFICATION OF CONSTRUCTED SYSTEMS 197

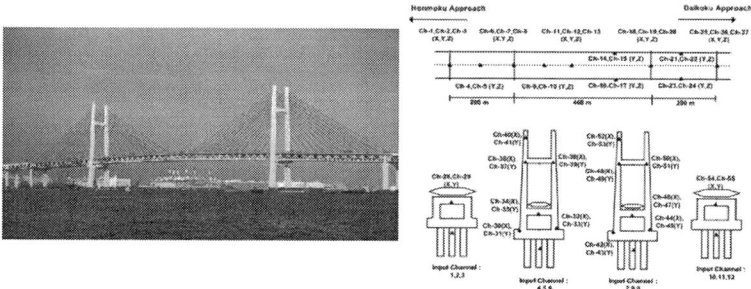

Figure B-26. Yokohama Bay Bridge and sensor layout for seismic-induced vibration monitoring

links with the length of 2m and the shape of an inverted pendulum were utilized to connect the girder with the towers and end piers.

B.6.2   Objective of Structural Identification Application

Because of the high intensity of seismic activities in Japan, monitoring for seismic response has been widely employed for decades especially for bridges with special features such as curved decks (Siringoringo and Fujino 2007) and bridges with new technologies such as base-isolation (Chaudhary et al. 2000). The Yokohama Bay Bridge was constructed on a soft soil that necessitated a new special foundation system (*The yokohama bay bridge*, 1991). It is also located near an active fault and close to the epicenter of the 1923 Great Kanto Earthquake. These conditions have made seismic performance a major concern. Therefore to confirm the seismic design and to monitor the bridge performance during earthquakes, a comprehensive and dense array monitoring system was installed. The objectives of the monitoring system are the evaluation of seismic performance, verification of and comparison with seismic design, and observation of possible damage. Particular attention is given to the seismic isolation device in the form of Link Bearing Connection (LBC).

B.6.3   System Identification Method Development

Seismic records with varying amplitude obtained from six major earthquakes from 1990 to 1997 were analyzed to evaluate global and local performance of the bridge (Siringoringo and Fujino 2006). System identification of the long-span bridge under seismic excitation requires that non-unique ground excitation records measured along the bridge and excitation in multiple directions be taken into account. Therefore, multi-input and multi-output system identification is adopted (Siringoringo and Fujino 2008a). The system identification procedure makes use of the correlation data between input-output to realize the state-space model and estimate the modal parameters.

To apply this system identification, an input-output relation should be firstly defined. In case of earthquake excitation, base motions can be considered as direct source of excitation and therefore utilized as inputs to the system so as to minimize the effect of soil-structure interaction. Structural responses of superstructure elements such as girder, piers and the towers are considered as the outputs of the system.

B.6.4   Experimental Studies

As part of a dynamic monitoring system, the bridge is equipped with 85 channels of accelerometers at 36 locations. These accelerometers have a frequency range between 0.05 to 35Hz with accuracy of 15 microampere per gals. Among these 85 channels, 25 are located on the substructure (such as pile foundation and pile caps) and the rest are installed on the superstructure (towers, pier caps and girder) (Figure B-27). Along the girder, sensors were installed at 9 locations with a space of 115 m between each. These sensors measure accelerations vertically, laterally (out-of-plane) and longitudinally (bridge axis). Both towers are equipped with accelerometers measuring longitudinal and lateral movement at both of the bridge's H-shaped columns. This enables the identification of pure longitudinal, lateral and torsional modes. For end-piers, accelerometers were installed on the pile cap and pier cap. Using the response from the sensors on the pier cap and the girder just above the pier cap, one can observe the behavior of link-bearings connecting the girder with the towers and with the end-piers.

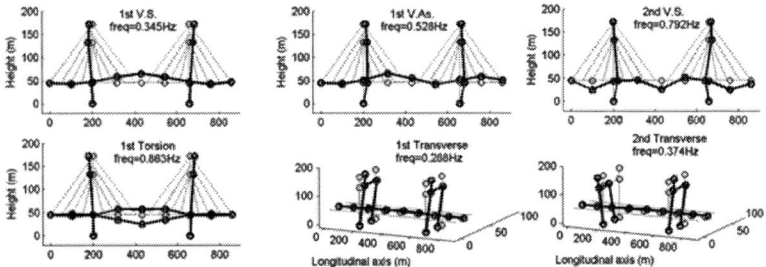

Figure B-27. Sensor layout of Yokohama Bridge and identified mode shapes

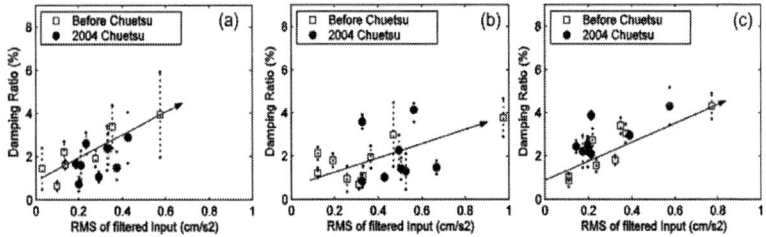

Figure B-28. Variation of damping ratio of Yokohama Bay Bridge with respect to earthquake input amplitude: (a) 1st vertical bending mode, (b) 2nd vertical bending mode (c) 1st transverse mode (Source: Siringoringo and Fujino 2008a, reproduced with permission from John Wiley and Sons)

STRUCTURAL IDENTIFICATION OF CONSTRUCTED SYSTEMS   199

B.6.5   Data Analysis and Interpretation

Because the system identification was applied to earthquake records of different amplitudes, the relationship between the identified characteristics and earthquake input intensity can be established. The analysis summarizes trends observed from the system identification results of the Yokohama Bay Bridge, whose identified mode shapes are shown in Figure B-27, as follows (Siringoringo and Fujino 2006):

- Natural frequencies for bending modes in vertical and lateral directions, as well as the torsional modes, are generally constant without significant changes with respect to the input magnitude. Some of the higher lateral modes such as the 5th and 6th modes were slightly decreased, but these decreases are not very significant (1 to 3%).

- The average damping ratios for higher vertical bending modes (3rd, 4th and 5th modes) were found within the range of 2-3% and show no significant changes with increased earthquake amplitude. The damping ratios for the last two torsional modes (2nd and 3rd modes) are also found in the range of 2-3% and mostly remain constant with increased earthquake amplitude. Average damping ratios for lateral modes, however, are higher than those of the vertical modes, with the range of average values 3-5%.

- Damping ratios of lower modes (1st and 2nd modes) in both vertical and lateral direction show an increasing trend with the increase of earthquake magnitude. For small magnitude, average damping ratios are found to be 2% and increase significantly up to 4-5% as the earthquake magnitude increases. The result indicates that damping ratios of lower modes are dependent on earthquake amplitude, which might be due to the greater energy dissipation caused by friction in bearings that occurs during large earthquake.

In general, natural frequencies identified from earthquake records are in good agreement with those from the ambient and forced vibration tests. The frequencies are almost constant with respect to earthquake amplitude. Modal damping ratios of several lower modes, however, indicate magnitude dependence, as shown in Figure B-28. Most damping ratios are identified with average values in the range of 0.5-5.5%. This lowest average value (0.5%) is satisfactory according to the minimum damping ratio required by the bridge seismic design and specification. The highest average value (5%), however, is much larger than the previously estimated 2% from the study conducted by the Honshu-Shikoku Bridge Authority.

Performance of local structural components can also be evaluated when measurement is conducted using dense arrays of sensors. An example of such evaluation is the assessment of Link Bearing Connection (LBC) using modal characteristics identified from strong motion records. LBC is a type of connection designed to minimize the inertial force of superstructure from being directly transferred to substructures. Investigation of the LBC (Figure B-29) is essential, considering the importance of its performance during earthquakes. In design, the LBC is expected to function as a hinged connection especially during large vibrations in longitudinal direction. This implies that the girder and pier-cap work as separated

200 STRUCTURAL IDENTIFICATION OF CONSTRUCTED SYSTEMS

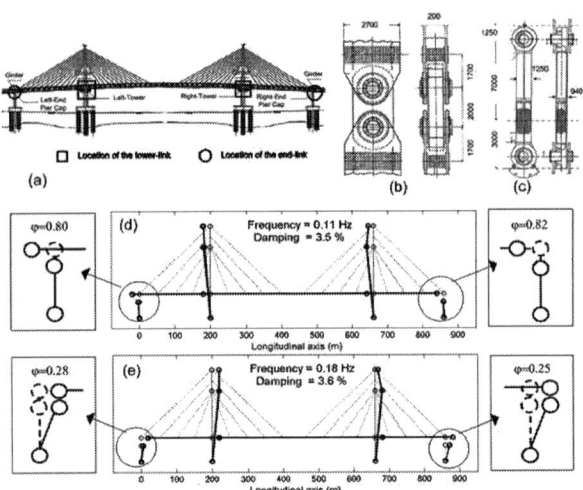

Figure B-29. (a) Location of link bearing connection of Yokohama Bay Bridge; (b) typical LBC at the tower, (c) typical LBC at end piers. Two of the three typical first modes of Yokohama Bay Bridge identified from the main shock at 17:57 (d) hinged-hinged mode (e) fixed-fixed mode

units, and therefore the force from the superstructure will not be transmitted into the end-piers.

Investigation of link-bearing performance of Yokohama Bay Bridge is carried out using the records from fourteen earthquake frames. The analysis involves system identification especially by observing the first longitudinal mode, analysis of the response between the pier-caps and the girder, and the analysis using finite element models. Fourteen frames of earthquakes from 1990 to 1997 were analyzed with the maximum input excitations varied from 2 to 14 cm/s^2. Based on the analysis, the following findings are observed:

- Three typical first longitudinal modes were found from system identification with the main focus on the relative modal displacement between end-piers and girder. They are: the hinged-hinged mode, mixed hinged-fixed mode and the fixed-fixed mode. The latter two modes are variations of what was highly expected mechanism (hinged-hinged mode). The response analysis of relative displacement between the end-piers and girder confirms these findings.

- During small earthquakes, the LBC has yet to function as a fully hinged connection. Therefore higher natural frequencies due to the stiffer connection were observed. The mixed hinged-fixed mode was observed during moderate earthquakes. The fully hinged connections at both of the end-piers were observed mostly during large earthquakes.

STRUCTURAL IDENTIFICATION OF CONSTRUCTED SYSTEMS   201

B.7   Alfred Zampa Memorial Bridge

B.7.1   Bridge Description and Experimental Studies

The Alfred Zampa Memorial Bridge (AZMB), a newly built long-span suspension bridge, is located 32km northeast of San Francisco on interstate Highway I-80. The AZMB is the third bridge crossing the Carquinez Strait and replaces the original bridge built in 1927 (Figure B-30). With a main span of 728m and side spans of 147m and 181m, the AZMB is the first major suspension bridge built in the United States since the 1960s. The design and construction of the AZMB incorporates several innovative features that have not been used previously for a suspension bridge in the USA, namely (1) orthotropic (aerodynamic) steel deck; (2) reinforced concrete towers; and (3) large-diameter drilled shaft foundations. The AZMB is also the first suspension bridge worldwide with concrete towers in a high seismic zone.

Figure B-30. Alfred Zampa Memorial Bridge
(Source: Courtesy of Leon Bacud. http://en.wikipedia.org/wiki/File:
Carquinez_Bridge_9-1-2007_From_Northeast.jpg)

202 STRUCTURAL IDENTIFICATION OF CONSTRUCTED SYSTEMS

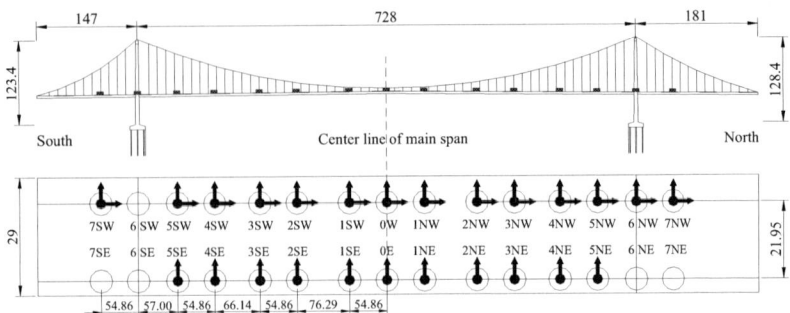

Figure B-31. Overall dimensions of the AZMB and instrumentation (accelerometers) layout (units: m)

B.7.2 System Identification Methods Used

In this study, three output-only system identification methods, namely the multiple-reference natural excitation technique combined with the eigensystem realization algorithm (MNExT-ERA), the data-driven stochastic subspace identification (SSI-DATA) method, and the enhanced frequency domain decomposition (EFDD), were applied to identify the modal parameters of the bridge based on bridge vibration data collected from two types of tests: ambient vibration test and forced vibration tests based on controlled-traffic loads.

B.7.3 Experimental Studies

A set of dynamic field tests were performed on the AZMB in November 2003, just prior to its opening to traffic. These tests included ambient vibration tests (mainly wind-induced) and forced vibration tests based on controlled traffic loads and vehicle-induced impact loads. The controlled traffic loads consisted of two heavy trucks (about 400kN each) traversing the bridge in well-defined relative positions and at specified velocities, while the impact loads were generated using one or both trucks driving over triangular shaped steel ramps (60cm long and 10cm high) designed and constructed specifically for these tests. Four traffic load patterns and seven vehicle-induced impact load configurations were used in the forced vibration tests. The vibration response of the bridge was measured through an array of 34 EpiSensors ES-U (uni-axial) and 10 EpiSensor ES-T (tri-axial) force-balance accelerometers from Kinemetrics Inc. installed at selected locations (stations) along both sides of the bridge deck covering the entire length of the bridge (Figure B-31). Along the west side of the bridge deck, 14 stations were instrumented with either a single EpiSensor ES-T or three EpiSensors ES-U at each station to measure the vertical, transversal and longitudinal motion components. The east side of the bridge deck was instrumented with 22 EpiSensors ES-U at 11 stations (i.e., two uni-axial accelerometers per station) measuring the vertical and transversal motion components. Instead of using roving accelerometers at different measurement stations with fixed accelerometers at one or more reference stations (as commonly done for dynamic testing of bridges), a total of

STRUCTURAL IDENTIFICATION OF CONSTRUCTED SYSTEMS   203

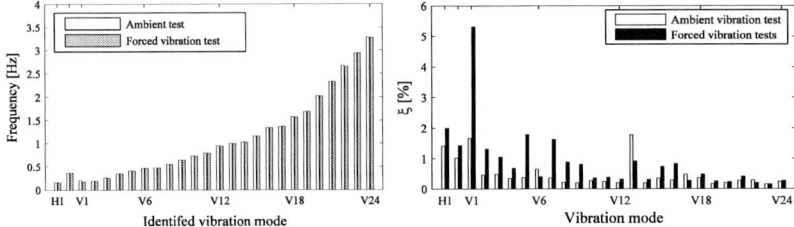

Figure B-32. Comparison of natural frequencies (left) and damping ratios (right) identified using ambient vibration and forced vibration test data

64 channels of acceleration response data were recorded simultaneously in the tests described above, consisting of 25 vertical, 25 horizontal, and 14 longitudinal motion components. These dynamic field tests provided a unique opportunity to determine the dynamic properties of the AZMB in its as-built (baseline) condition with no previous traffic loads or seismic excitation. More details about the bridge and the dynamic tests performed can be found in Conte et al. (2008).

B.7.4   Data Analysis and Interpretation

From the modal identification results obtained (He et al. 2009), it was observed that the natural frequencies identified using the three different methods are in excellent agreement, while the relative difference in the damping ratios identified using different methods is significantly larger. This is a well known fact widely reported in the structural identification literature, namely that the estimation uncertainty of the damping ratios is inherently higher (by more than an order of magnitude in the coefficient of variation) than that of the corresponding natural frequencies. Figure B-32 shows the average (over three methods) identified natural frequencies and damping ratios of the first two horizontal and 24 vertical (or torsional) modes. From these figures, it can be seen that:

- The natural frequencies identified based on data from the two different types of tests are in excellent agreement, except for the 1-AS-V (first anti-symmetric vertical) mode. The significant difference in the identified natural frequencies for this mode reflects the difficulty in identifying it due to its very low relative contribution to the measured bridge vibration in both the ambient and forced vibration tests;

- For most vibration modes, especially for the lower vibration modes, the averaged modal damping ratios identified using forced vibration data are higher than those identified using ambient vibration data.

A 3D representation of the normalized mode shapes for the first 15 identified modes is given in Figure B-34. The identified space-discrete mode shapes were interpolated between the sensor locations using cubic splines along both sides of the bridge deck and straight lines along the deck transverse direction. Finally, the identified natural frequencies and mode shapes were compared with their analytically

204 STRUCTURAL IDENTIFICATION OF CONSTRUCTED SYSTEMS

predicted counterparts obtained from a 3D finite element model used in the design phase of the AZMB and not modified a posteriori to better fit the identified modal parameters. These are found to be in good agreement for the modes contributing significantly to the measured response (Figure B-33).

The system identification results obtained from this study provide benchmark modal properties of the AZMB, which can be used as a baseline in future health monitoring studies of this bridge. From the facts that (a) very different methods provide similar results for the modal parameters of the modes contributing most to the measured bridge vibration, (b) the natural frequencies and mode shapes identified using two different types of test data are in good agreement, and (c) these methods were found in a recent study (Moaveni et al. 2007) to provide modal parameter estimates with low bias and variability for the natural frequencies and mode shapes, it can be concluded that it is likely that the identified natural frequencies and mode shapes are close to the actual modal parameters of this bridge. Although the damping ratio estimates provided by this study have a much larger variability across methods (than the natural frequencies and mode shapes), the average values over the three methods are likely to be representative of the actual effective damping ratios of the bridge at the two response levels considered.

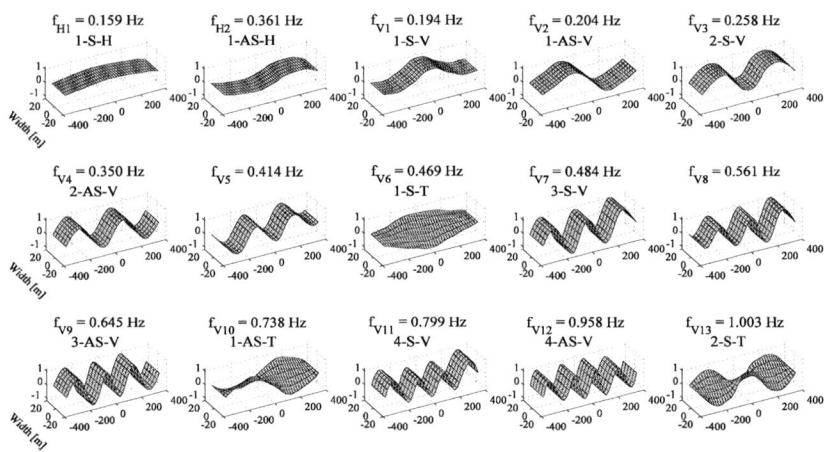

Figure B-33. Normalized vibration mode shapes identified using MNExT-ERA based on ambient vibration data (S = Symmetric; AS: = Anti-Symmetric; H, V, T = Horizontal, Vertical, and Torsional mode, respectively)

Overall, all three system identification methods applied in this study performed very well in both types of tests. However, the use of several system identification methods is recommended for cross-validation purposes and for avoiding missing modes, as different methods provide modal parameter estimators with different intra-method and inter-method statistical properties (bias, variance,

STRUCTURAL IDENTIFICATION OF CONSTRUCTED SYSTEMS 205

covariance), which depend on the frequency content of the input excitation and the level of violation of the assumed stationarity. It should be noted that the performance of the EFDD method is not as robust as that of the other two methods, since it requires user intervention for peak picking in the identification process.

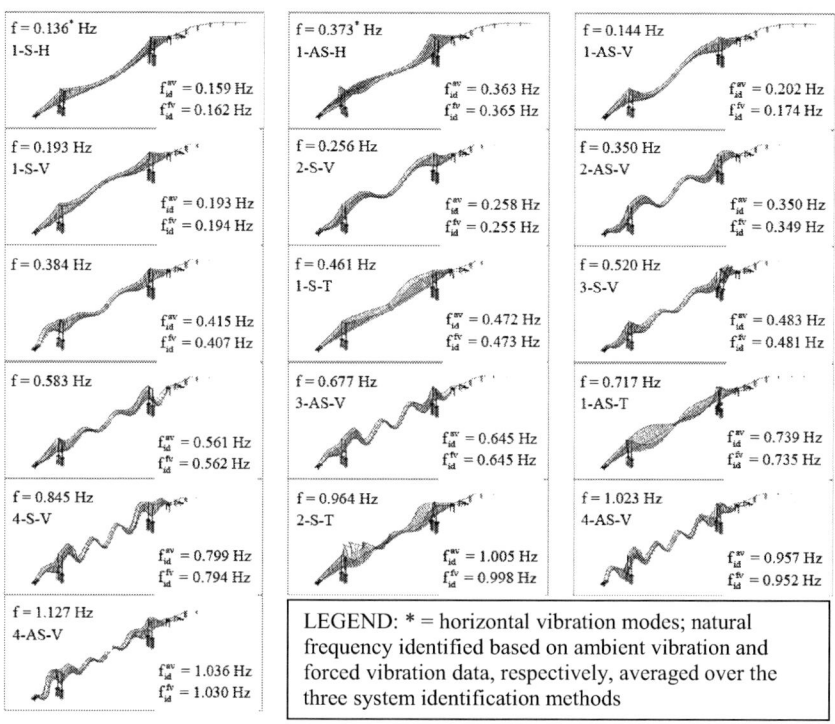

Figure B-34. Vibration mode shapes of the AZMB computed from the bridge finite element model in ADINA

B.8 Langensand Bridge

B.8.1 Bridge Description

The new Langensand Bridge in Lucerne (Switzerland) is a steel-concrete composite girder with a 80m long span. The slenderness ratio (L/h) of the girder varies from approximately 55 at the abutment to 30 at mid-span. Figure B-35 shows the main girder profile and its boundary conditions. The structure is being built in two phases to avoid traffic interruption on the existing bridge. Load tests were performed after the completion of the first phase when only a half of the bridge was completed (Figure B-36).

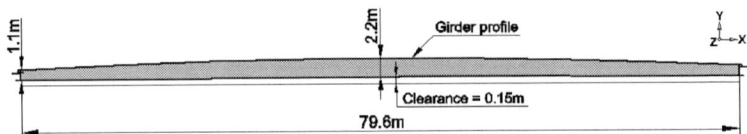

Figure B-35. Elevation profile of the main girder of Langensand Bridge

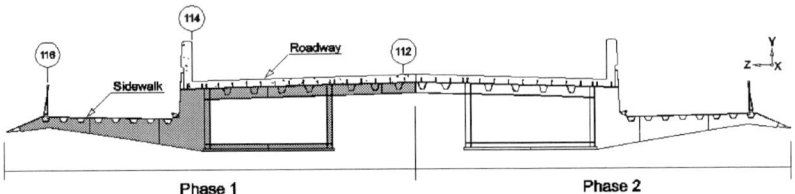

Figure B-36. Two construction phases of Langensand Bridge

B.8.2 Objectives of Structural Identification

The Langensand Bridge is a good example of innovative structural engineering using composite design. During the design stage, engineers at Guscetti & Tournier SA made justifiably conservative assumptions regarding aspects such as the composite behavior and support conditions. Behavioral models using these assumptions often underestimate the load-bearing capacity of the structure. An accurate estimate of the reserve capacity of the bridge was sought by the owner (City of Lucerne) who manages many structures in the city. Such an estimate is useful to the owner for tasks such as routing heavy vehicles across the city and for later decisions in cases of modifications and deterioration. Therefore, the identification task was to verify the design hypothesis that the bridge achieves full composite action under service loading. Specific objectives of the study were as follows, with results available in the paper by Goulet et al. (2010):

- to estimate the as built in-service reserve capacity of the bridge beyond design limits;
- investigate the applicability of the multi-model approach presented in Section 5.1.1.2 on full-scale structures;
- compare the results with those from single-model updating ignoring errors;
- find an accurate behavioral model of the bridge for subsequent static and dynamic; analyses and additionally, to help the owner manage the structure more efficiently.

B.8.3 Model Development

Figure B-37 shows the detailed cross-section of the finite element model of Langensand Bridge generated in ANSYS (ANSYS 2007). This model has several parameters that are assigned values by stochastic sampling during the candidate model generation process. The model incorporates elements of the bridge that contribute to the stiffness of the structure such as barriers.

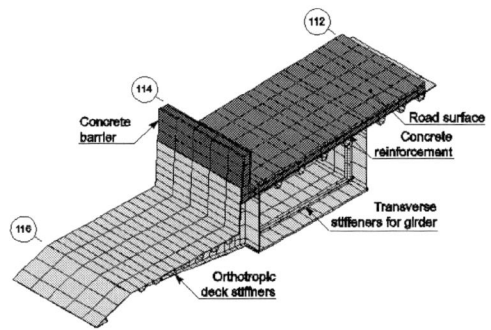

Figure B-37. Cross-section of Langensand Bridge finite element model

B.8.4 Experimental Studies

The following types of measurements are taken during the static-load test, as shown in Figure B-38:

- displacement measurements taken at six locations (at the crossing of the axis: S7-112, S7-116, S12-112, S12-116, S17-112 and S17-116) with optical devices;
- rotations about the z-axis are measured using two inclinometers placed near the abutment (at right of the axis: A1-112 and S7-112); and
- deformations are measured at three locations on the bridge along the section S13 using fiber-optic sensors. Two sensors are placed along the X-direction at the centre of the steel girder such that one is embedded in the top chord and the other on the bottom chord of the concrete deck. The third sensor is placed on the bottom chord of the steel girder.

Sources of errors introduced during the modeling and measurement process were identified. The contribution of each source was also quantified, as summarized in Goulet et al. (2010).

208   STRUCTURAL IDENTIFICATION OF CONSTRUCTED SYSTEMS

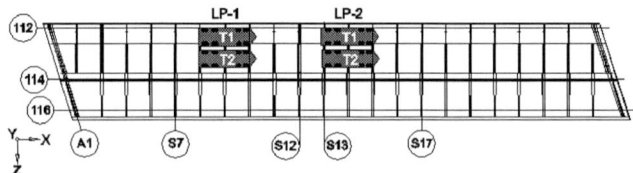

Figure B-38. Top view of Langensand Bridge with reference axes for measurement locations

B.8.5   Data Analysis and Structural Identification

The multi-model candidate selection approach described in Section 5.1.1.2 is used for structural system identification. A set of 1000 models is generated by sampling several modeling assumptions. Models are then filtered using collected measurements to obtain the set of candidate models. The process and the resulting set of candidate models are presented in Figure B-39. This figure shows that the candidate models, which are representative of the structure's measured behavior, do not minimize the discrepancy between measurements and predictions. These results indicate that the set of candidate models predict the measured displacement and rotation within 4% to 7%. Strains are more difficult to assess. The discrepancy in strains ranges from 15 to 23%.

These results illustrate that, when errors are explicitly considered within the systematic framework, biasing the identification toward an incorrect model is avoided. The candidate models were able to predict the service behavior of the structure to within 7% of measured values. Estimates using candidate models showed that the structure has 30% reserve capacity compared with the design model for the vertical displacement criteria. The study also found that minimizing the discrepancies between predictions and measurements would have lead to wrong models that overestimate the reserve capacity. Finally, the results revealed that the bridge was behaving in a fully composite manner for the tested loads.

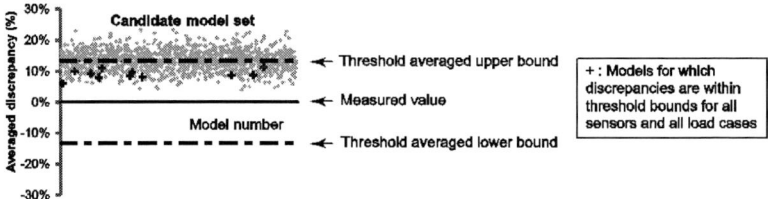

Figure B-39. Candidate model selection process using threshold

STRUCTURAL IDENTIFICATION OF CONSTRUCTED SYSTEMS 209

B.9 Sunrise Boulevard Movable Bridge

B.9.1 Bridge Description

A movable bridge is a structure that has been designed to have two alternative positions and can be moved back and forth between those positions in a controlled manner as a way for land traffic to cross a waterway while ensuring a path for the waterborne traffic (Koglin 2003; Wallner and Pircher 2007). Florida has a large population of movable bridges due to the waterways and coastal topography. Most of these bridges are owned by the Florida Department of Transportation (FDOT). The FDOT has an inventory of 98 movable bridges including 3 lift type, 94 bascule type, and 1 swing type. The Sunrise Movable Bridge is a bascule bridge over the Florida Inter Coastal water way (Figure B-40). This span was constructed in 1989. It has double bascule leaves, each 22.49m long approximately, and 26.15 m wide, carrying

Figure B-40. Sunrise Boulevard Bridge in Florida

three traffic lanes and opening about 15 times a day. The bascule leaves are lifted horizontally at the point of the trunnions, which are the pivot points on the main girders. The weight of the span is balanced with a counterweight that minimizes the required torque to lift the leaf. In the closed position, the girder rests on a support referred as Live Load Shoe (LLS) on the pier and traffic loads are not transferred to the mechanical system. The movable bridge also involves fixed components, such as piers and approach spans.

B.9.2 Objective of Structural Identification Application

Movable bridges experience major deterioration as compared to regular fixed bridges due to their complex structural, mechanical and electrical systems (Catbas et al. 2007) and even a minor malfunction of any component can cause an unexpected failure of bridge operation (Buxton-Tetteh 2004). This necessitated a comprehensive SHM plan to monitor for the most common issues associated with movable bridges. This includes observations of traffic-induced strains and accelerations and their comparison with analytical results coming from a Finite Element Model. Strain data from opening and closing operations are also evaluated and compared with FE model results. These comparisons are used to verify the FE model, and the Inventory and Operating Load Rating is evaluated.

B.9.3 Experimental Studies

As a part of the ongoing research project for FDOT, main issues for the maintenance of electrical, mechanical and structural components of the movable bridge were identified. Based on these, an extensive sensor network is designed and implemented to monitor various parts of the bridge. A total of 168 sensors are deployed to the bridge for monitoring the electrical, mechanical and structural components as well as collecting environmental data. The electrical and mechanical components are monitored with accelerometers, strain rosettes, tiltmeters, microphones, infrared temperature sensors, ampmeters, video cameras, and pressure gages. Structural components are mainly monitored with accelerometers, high-speed strain gages and slow speed vibrating wire strain gages. A video camera detects the traffic and relates it to the other measurements. Finally, a weather station measures wind speed, wind direction, humidity, temperature, barometric pressure, and rain. Figure B-41 shows the installed monitoring system and the sensor locations at the main girder. The structural parts of the bridge are investigated by using the data from the accelerometers and strain gages at the main girders.

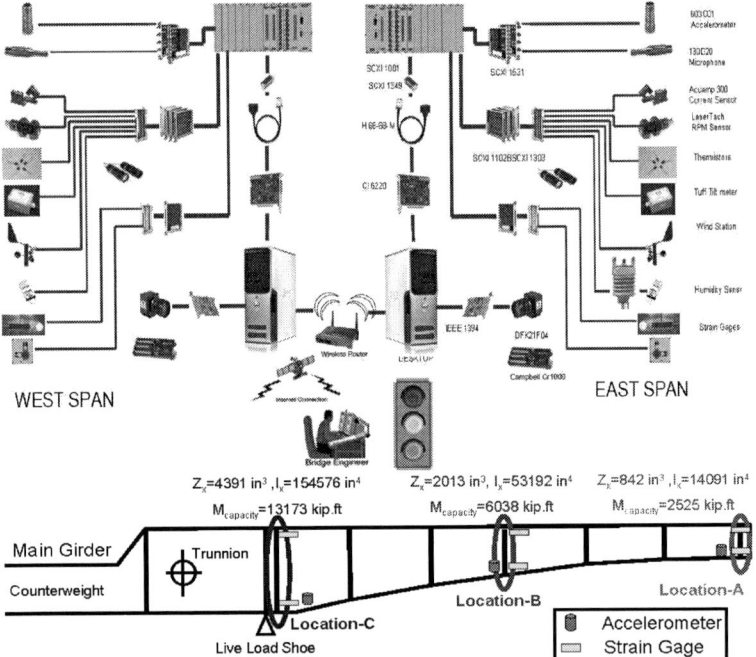

Figure B-41. Sensor network used in the movable bridge project (top) and sensor locations at the main girder (bottom)

STRUCTURAL IDENTIFICATION OF CONSTRUCTED SYSTEMS   211

B.9.4   System Identification Method Development

The analytical model used in this study was developed using a finite element analysis program. Construction plans and details of the Sunrise Bridge were closely studied to ensure a proper modeling of the superstructure. There are a few main components of the bridge superstructure that are critical to accurately model the local behavior of the deck and secondary beams as well as the global behavior. The first main component of the bridge was the main girders where boundary conditions were imposed at the trunnion and live load shoe locations. The second main component of the system to be created was the floor beams, sidewalk brackets, and diagonal bracing which were composed of frame elements. These elements were connected to main girders and each other with rigid links at the centroids. Once all secondary beams were created, the deck of the bridge was constructed. The deck was modeled using 4-node quadrilateral shell elements and connected to the main girders and secondary beams using rigid links. Finally, solid elements were created to model the concrete counterweight. The model and its parameters can be seen in Figure B-42.

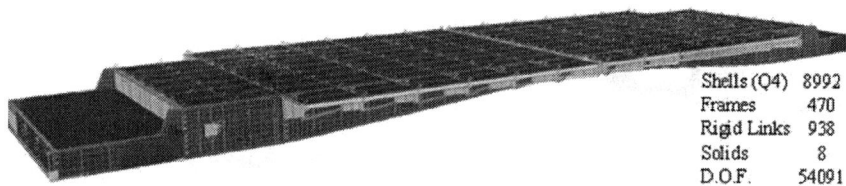

Shells (Q4) 8992
Frames   470
Rigid Links 938
Solids   8
D.O.F.   54091

Figure B-42. Final FEM of Sunrise Bridge

B.9.5   Data Analysis and Model Calibration

Ambient acceleration data were collected by using 16 sensors that are located at critical locations of the bridge in both vertical and horizontal directions. Based on the preliminary FEM analysis, these sensors can adequately capture the dynamic behavior of the bridge. A sample data set is shown in Figure B-43. In this study, a Complex Mode Indicator Function (CMIF) based modal parameter estimation technique is used along with the Random Decrement (RD) technique. First, the ambient vibration data is averaged by using RD to obtain unscaled free response data. Then, the modal parameters are identified with CMIF using the unscaled free responses. A detailed discussion about the methodology is beyond the scope of this study and more information can be found in Gul and Catbas (2008).

212   STRUCTURAL IDENTIFICATION OF CONSTRUCTED SYSTEMS

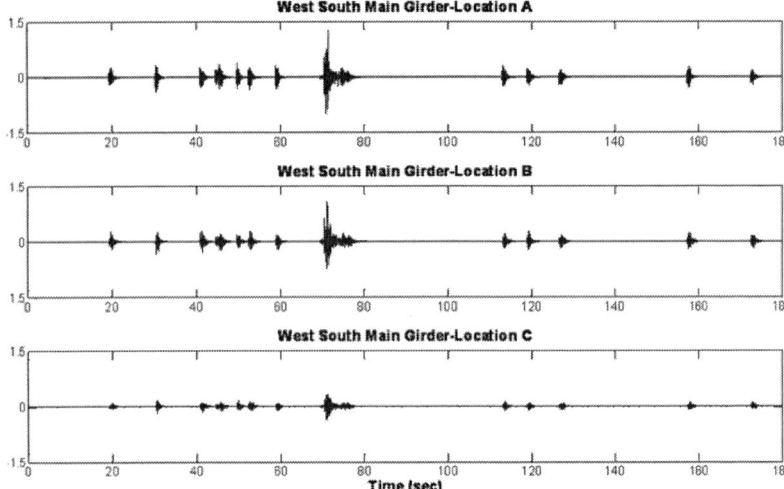

Figure B-43. Ambient acceleration data from the west south main girder of the bridge

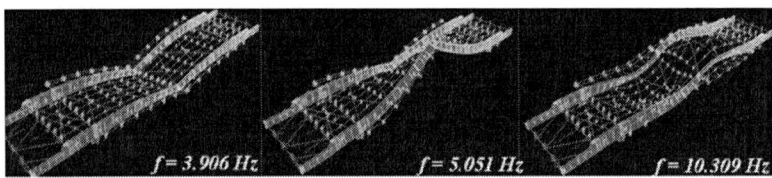

Figure B-44. Experimental and analytical frequencies model, and the first three FEM mode shapes

The modal parameters identified from the field data are presented in Table B-3 along with the first three mode shapes of the FEM in Figure B-44. The comparison of the dynamic results shows that the FEM can capture the global behavior of the structure quite satisfactorily. The model can be calibrated and further improved to obtain a better match with the field data; however, this is not investigated in the current phase of the study as the current fit is deemed quite satisfactory. These results also verify the consistency of the field data, although it is noted that the data quality can be improved with future investigations with higher resolution sensors, higher dynamic range data acquisition systems with dedicated A/D converters, improved cabling, connections etc.

STRUCTURAL IDENTIFICATION OF CONSTRUCTED SYSTEMS 213

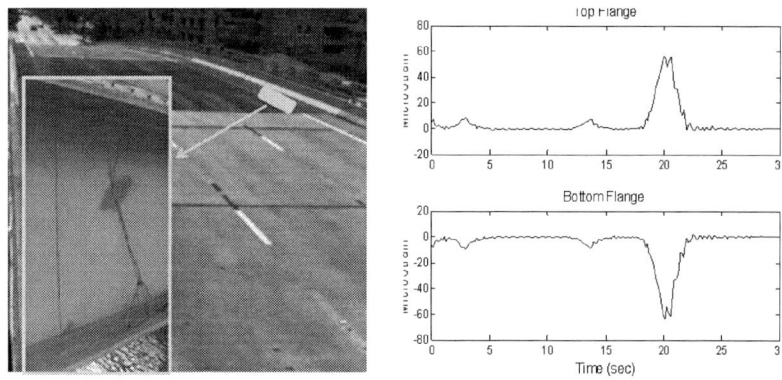

Figure B-45. Location of the sensor (left) and measurements of the strain (right)

Table B-3. Comparison of the first three natural frequencies

Mode #	Field Data (Hz)	FEM (Hz)
1	3.543	3.906
2	5.132	5.051
3	9.286	10.309

Traffic-induced strain data was collected at different locations including the live load shoes, span locks, floor beams and different sections of the main girders. Here the data coming from the live load shoes (support area) is investigated since the effect of the traffic loading is higher in these locations. Location of this sensor and sample strain measurements are illustrated in Figure B-45.

As it is observed from the plots in Figure B-45, the strain values at around 20 seconds are considerably higher than the rest of the values. After noting these high strain values, the recorded video images were investigated. It is determined that this increase in the strain was caused by a Riverside Transit Agency (RTA) bus crossing the bridge (RTA 2004). The properties of the bus were identified and they were incorporated in the FEM so that the RTA bus could be simulated by applying the point loads corresponding to the wheels of the bus over the bridge as illustrated in Figure B-46. A good consistency between the response of the FEM and real structure is observed. The maximum response developed for the upper flange is 56 $\mu\varepsilon$ (microstrains), while the calculated strain in FEM for the same flange is 53 $\mu\varepsilon$. Similarly, the maximum response developed for the lower flange is -63 $\mu\varepsilon$, while the calculated strain in FEM for the same flange is -57 $\mu\varepsilon$. The difference between the measured and calculated strain for upper flange is 5 % and for the lower flange is 10%. These results indicate that generated FEM is in good agreement with actual bridge data for localized strain measurements as well.

214 STRUCTURAL IDENTIFICATION OF CONSTRUCTED SYSTEMS

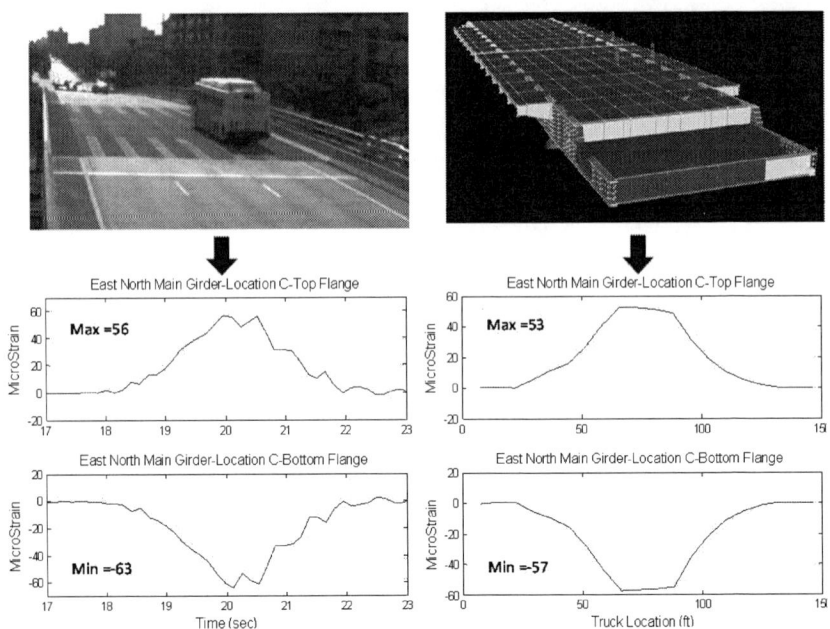

Figure B-46. The effect of typical RTA transit bus experimental (left) and FE model (right)

As another local behavior comparison, the strains developed at the live load shoe area during opening and closing are compared for field and FEM data Figure B-47 shows data collected during the two successive opening and closings of the bridge at the East South side. As shown in this figure, the strain variation for upper flange is 164 µε whereas it is 127 µε for lower flange. Then, opening and closing were simulated in FEM and strain developed for each case was also recorded as shown in Figure B-47. The strain variation for the upper flange is 138 µε, which is 16% different from the experimental data while for the lower flange it is 132 µε which is 4% different from the experimental data. This again indicates that FE model agrees well with experimental data.

B.9.6 Interpretation and Decision Making

The load rating of a bridge can be expressed as the factor of the critical live load effect to the available capacity for a certain limit state. Since the live load can play a critical role in the distribution of loads on a short to medium size bridge, load rating analysis is commonly used as an effective approach to evaluate live load carrying capacity of bridges. Load rating can be carried out for a number of critical locations and components of the structure. In this part of the study, the load rating of the movable bridge was calculated by following the AASHTO guide (AASHTO 2004) and using the FEM, which was verified using global and local responses measured at the bridge.

For this study, a moving truck load (HL-93) was simulated in the FEM and the live load moments were obtained for each truck location. Main girders were selected for load rating instead of floor beams because main girders are more critical from the structural system perspective. The transverse cross-section of the deck (Figure B-48) shows the location of the lanes, design lane load and axle position for the HL-93 standard truck loading. Since a 3D FE model was used, axle loads were defined as individual point loads and lane loads were defined as a distributed load equivalent to 0.64 kip/ft as mentioned in the AASHTO code. The yield strength of the steel is taken as 248 MPa, and the dynamic impact factor is taken as 33% for both inventory and operating ratings.

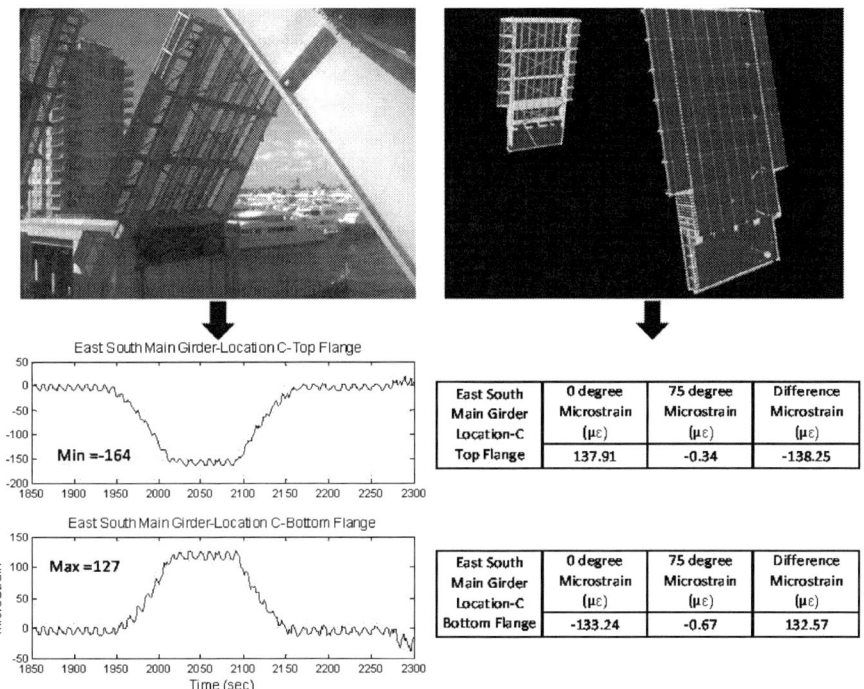

Figure B-47. Strains at East South live load shoe: measured responses between the 1860 second and 2300 second, where the bridge was opened and closed (left) and comparison to FEM simulation (right)

216 STRUCTURAL IDENTIFICATION OF CONSTRUCTED SYSTEMS

The load ratings based on the HL-93 truck on the East South main girder were calculated at the live load shoe, the middle of the leaf, and the tip, which are shown in Figure B-41. The capacities of the sections were calculated based on the ultimate moment capacity, which can be obtained by multiplication of the yield strength (F_y) and plastic section modulus (Z_x). It is seen that the cross-section at the live load shoe (Location-C) is more critical than the sections at Location-A and Location-B due to cantilever type configuration of the movable bridges. One important point here is that the critical truck configuration may not be easy to locate since the main girder is tapered as shown in Figure B-41. As a result, a moving load simulation is also conducted to determine the load rating at the most critical location as function of truck location. The most critical load rating was observed at Location-C when truck's front axle is at 84 ft, as shown in Figure B-48. The critical load rating values for each location can be seen in Table B-4.

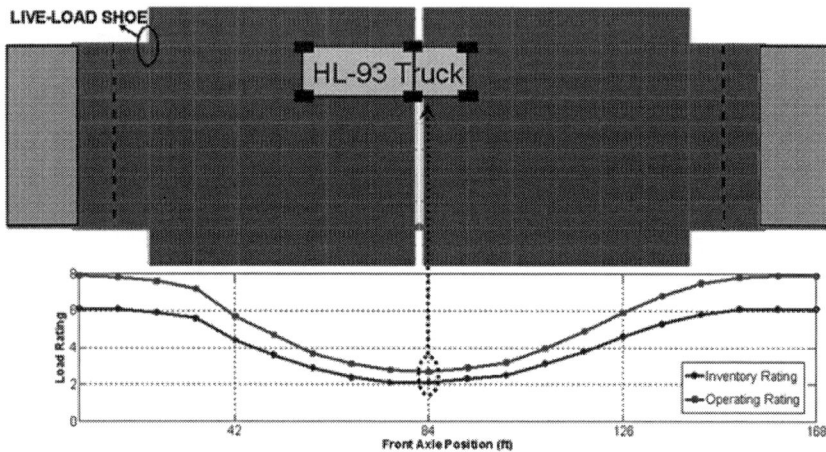

Figure B-48. Load rating values due to truck simulation for the live load shoe area

Table B-4. Load rating values for the three locations

Critical Load Rating Values	Inventory Load Rating	Operating Load Rating
Location-A	11.8	15.3
Location-B	3.9	5.0
Location-C	2.2	2.9

The rating at the most critical location for the most critical load configuration is observed to be acceptable and well above unity indicating that the bridge is able to carry this defined load. However, it should be indicated that it is more desirable to have an automated system that can continuously generate these load ratings based on the models and the monitored data. Recent research on movable bridge structural identification, structural health monitoring and evaluation by means of novel

STRUCTURAL IDENTIFICATION OF CONSTRUCTED SYSTEMS 217

technologies and methods has shown promising developments (Catbas et al, 2012a; 2012b; 2012c).

B.10 New Svinesund Bridge

B.10.1 Bridge Description

The new Svinesund Bridge in Figure B-49 is part of the European freeway E6 between Gothenburg and Oslo. It connects the Scandinavian countries Sweden and Norway over the Ide fjord and was opened for traffic in June 2005. With a total length of 704 m and a main span of 247 m, it is one of the longest single-arch bridges in the world (Figure B-50).

Figure B-49. Picture of the new Svinesund Bridge

The two steel box girders carry two lanes of traffic each and are 11.00 m wide. With a clear spacing of 6.20 m, they pass the concrete arch on either side. In the main span the girders are suspended from the arch by 6 hanger pairs, while in the side spans the girders are supported by reinforced concrete columns. Due to the slenderness of the arch, the bridge deck girders were designed to be rigidly connected to the arch in order to provide additional lateral stability. Due to the wide spacing of the bridge deck girders and the slender columns without column heads, it was necessary to prestress the bridge deck girders onto the columns to avoid uplifting for asymmetric loading. A more detailed description of the bridge is given in Darholm et al. (2007).

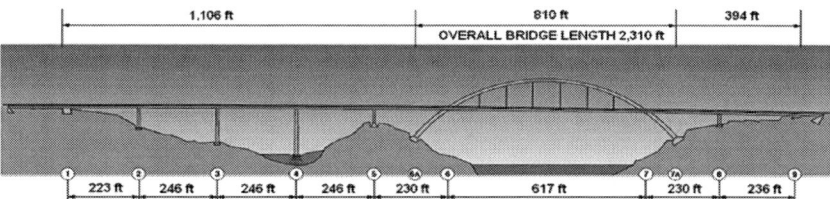

Figure B-50. Elevation of the new Svinesund Bridge (Source: Darholm et al, 2007; reproduced by permission from the Swedish Transport Administration)

B.10.2 Objective of St-Id Application

Due to the uniqueness of the design, a monitoring system was installed on the bridge to check that the bridge is built as designed, to get a better understanding of the bridge's behavior, and to produce an initial database (foot print) of the undamaged bridge (Karoumi and Andersson 2007). The monitoring program started during the construction phase and was kept running during the first years of service. Later the decision was taken to use the available measurement data for St-Id. The aim was to obtain a more accurate FE model, which may later, if kept up-to-date, be used for the structural assessment and maintenance. Furthermore, the possibility to use the measurement program to estimate uncertain structural parameters regarding boundary conditions and interactions between the parts of the structure was studied.

B.10.3 Model Development

The FE model had initially been developed for the design of the bridge. Hence, assumptions "on the safe side" have been appropriate and the level of detailing has been chosen for the design purpose and not for St-Id. Timoshenko beam elements have been used for the arch, while a grid of beams has been used to model the bridge deck girder (Figure B-51). The model was later converted into the FE software ABAQUS (Plos and Movaffaghi 2004). Jonsson and Johnson (2007) coupled the ABAQUS model to MATLAB as shown in Figure B-52. This allowed one to update model parameters, compute the corresponding response and calculate the objective function automatically, which has been used for St-Id in several studies (Jonsson and Johnson 2007; Schlune et al. 2008; Schlune et al. 2009).

Figure B-51. FE model of the new Svinesund Bridge

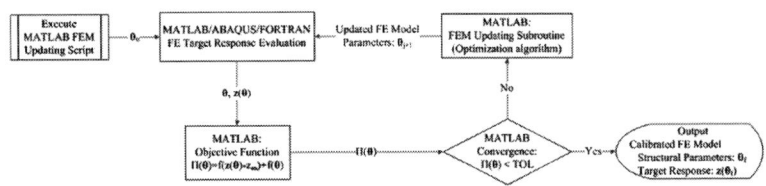

Figure B-52. Flowchart for coupling between MATLAB, ABAQUS and FORTRAN functions (Source: Jonsson & Johnson, 2007; reproduced by permission from D. Johnson)

STRUCTURAL IDENTIFICATION OF CONSTRUCTED SYSTEMS 219

B.10.4 Experimental Studies

The monitoring system and the load testing program focused on the response of the arch. A detailed description of the sensor types and sensor locations is presented in James and Karoumi (2003). In total, the arch has been equipped with 16 vibrating-wire strain gauges, 8 resistance strain gauges and 24 temperature sensors, which were cast into the concrete during construction. The strain sensors have been placed in sections near the arch foundation, in the arch underneath the bridge deck girders and at the arch crown. At least 4 sensors have been used for each section. In addition, one hanger pair has been equipped with load cells in order to monitor the hanger forces. The dynamic response of the bridge is monitored by 10 linear servo accelerometers. Furthermore, the lateral displacements of the bridge deck girder over supports 5 and 8 are monitored continuously by linear variable differential transformers. The weather is monitored by one outside temperature gauge and 3-directional ultrasonic anemometers which allow determination of the wind speed and wind direction. These permanently installed sensors have been used to study the effect of traffic, wind, and temperature on the bridge's response, to document changes in strains and dynamic properties, and to check the transverse movement of the deck over supports 5 and 8 (Ülker-Kaustell and Karoumi 2006).

Before the bridge was opened for traffic, it was instrumented with additional temporary sensors to allow for additional measurements under two days of load testing. Temporary accelerometers were installed on the hangers to allow estimating the hanger forces under dead weight by means of their identified eigenfrequencies. Extensometers and six total stations were used to collect additional data during static load testing with eight trucks. The extensometers were installed on the hangers and the total stations measured displacements of the bridge in 30 locations. In total, 5 different load patterns have been tested and the monitoring system and the temporary installed sensors have been used to measure the response of the bridge. Each load pattern has been repeated three times during the day with regular unloading of the bridge to allow removing the environmental effects from the measurements. Finally, trucks with leaf suspensions and air suspension were driven over the bridge with different velocities to determine the variation of the dynamic amplification factor with speed. The load testing program and results are described in more detailed in Karoumi and Andersson (2007).

B.10.5 Data Analysis, Model Calibration

Before the measurements from the load test could be used for updating, temperature effects have been removed from the measurements by linear interpolation (Figure B-53). Furthermore, Ülker-Kaustell and Karoumi (2006) analyzed 66 10-minute raw data files from ambient vibration monitoring by using the Maximum Likelihood Technique (MLT), the Random Decrement Technique (RDT) and the Stochastic Subspace Identification Technique (SSI). The identified eigenfrequencies and damping ratios together with the weather recordings have then been used to study the effect of the temperature and wind on the dynamic behavior of the bridge. It could be seen that the eigenfrequencies increase with decreasing temperature.

Before some model parameters were calibrated using an optimization algorithm, manual model refinements were introduced into the initial model. At first, the Young's modulus of the concrete arch was increased. This was necessary as the arch stiffness was based on the 28 days age of the plain concrete, while the arch was more than 1 year old when tested. By including the further hardening of the concrete and the contribution of the reinforcement bars, the arch stiffness was increased between 4-17% depending on the reinforcement degree.

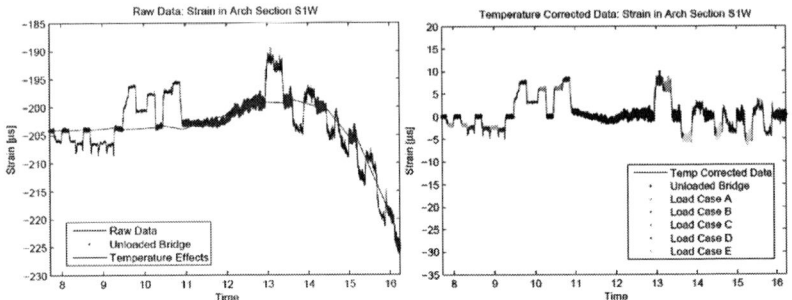

Figure B-53. Measured strains near southern arch foundation, before and after removal of temperature effects (Source: Jonsson & Johnson, 2007; reproduced by permission from D. Johnson)

Due to a significant disagreement of the second eigenfrequency of the initial FE model and the experimental counterpart ($f_{numerical}$= 0.460 Hz, $f_{experimental}$ = 0.846 Hz), Ülker-Kaustell and Karoumi (2006) tested to restrain the bridge deck girder movement over the columns in the model and obtained a better agreement. When this was introduced into the FE model the second eigenfrequency jumped to $f_{numerical}$= 0.885 Hz. However, at the same time the agreement for some tested truck load pattern decreased. This revealed a different bearing behavior under ambient vibrations and during the truck tests. Hence, the assumption of a restrained bridge deck movement was kept for the eigenfrequency calculations while a non-linear bearing response was assumed for the truckload tests. A linear-elastic behavior up to the static friction threshold and a displacement of 1 mm was assumed followed by a constant friction force for higher displacements (Figure B-54a).

The numerical hanger forces under dead weight were lower than the measured hanger forces (Figure B-54b). This showed that the mass of the non–structural parts was underestimated in the initial model. After introducing the mass of the asphalt layer, railing system and other non-structural parts via mass points into the model good agreement for the hanger forces in the middle of the bridge was obtained, while the hanger forces at the outer hangers were still underestimated. This might be a result of the construction process, which was beyond the scope of this study.

After these manual model calibrations, a parameter study was done by plotting an index of discrepancy between the numerical and experimental responses over model parameter variations. This allowed one to get an overview of the sensitivity of

STRUCTURAL IDENTIFICATION OF CONSTRUCTED SYSTEMS   221

model parameter changes to the measurement program. For sensitive model parameters, this further allowed one to obtain improved estimates of the actual structural parameters. The most significant change that was introduced into the model based on the parameter study was a stiffness increase of the bridge deck girders by 15%.

After manual calibrations, the model was fine–tuned by the use of an optimization algorithm. Tests of optimization algorithms for model calibration of a simple beam model using simulated measurements showed that the Gauss-Newton and other gradient–based optimization algorithms became unstable when artificial noise was added to the measurements. The Nelder-Mead Simplex algorithm required more model evaluations, but at the same time it was more stable for noisy measurements. Since a considerable amount of noise could be expected for the actual measurements, and since each model evaluation was not very time–consuming, the Nelder-Mead Simplex algorithm, implemented as 'fminsearch' in MATLAB, was used for fine–tuning of the model. The parameters that were fine–tuned are the E-modulus of the concrete arch, the E-modulus of the bridge deck girders, the additional mass of non-structural parts, and the static friction threshold of the bearings. To allow summing up the residuals of the different measurement types into one objective function, the residuals have been normalized by the standard deviation of the corresponding measurement types.

The accuracies of the initial and the calibrated model are shown in Shulune et al, (2008) by plotting the numerical responses over the experimental counterparts. Good agreement between numerical and experimental responses is obtained when markers are close to the line of equality. It can be seen that the model accuracy could significantly be improved for the eigenfrequencies, displacements and hanger forces. For the strains only a slight improvement could be achieved. Plausible reasons are the still quite simplified modeled bearing response, a too simplified FE model, or measurement errors. A more detailed description of the model calibration process can be found in Schlune et al. (2009) and Jonsson and Johnson (2007).

B.10.6   Interpretation and Decision Making

In this case study St-Id was applied to a newly constructed bridge and it was shown

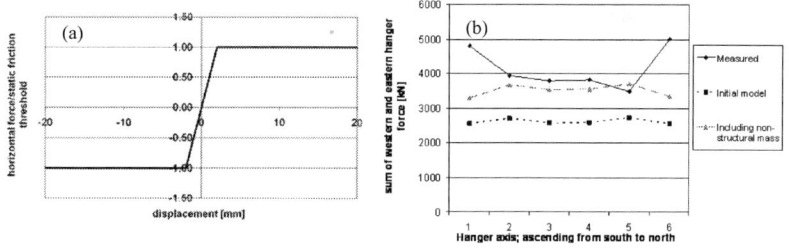

Figure B-54. (a) Load displacement curve for non-linear bearings; (b) Comparison of experimental and numerical hanger forces (Source: Schlune et. al. 2008; reproduced with permission from CRC Press)

that the bridge behaves as designed. The high discrepancy in the numerical and experimental eigenfrequencies could be explained by a restrained bridge deck movement under ambient vibration. Furthermore, the load tests revealed a stiffer behavior of the bridge. This could partially be explained by the increased arch stiffness due to further hardening of the concrete and the contribution of the reinforcement to the arch stiffness. Besides that, an increase of the bridge deck girder stiffness by about 15% led to best agreement between the experimental and numerical responses. The reason for this is still unclear but it might be due to the contribution of the railing system.

Furthermore, a high influence of temperature and other environmental conditions on the monitored properties could be observed. Hence, an improved understanding of structural changes due to temperature and environmental conditions is needed to discern structural changes due to temperature and environmental conditions form structural changes due to damage.

B.11 References

AASHTO. (2004). *LRFD bridge design specifications,* American Society of Highway Transportation Officials, Washington, DC.

Abdel-Ghaffar, A. M., and Rubin, L. I. (1983a). "Lateral earthquake response of suspension bridges." *ASCE Journal of the Structural Division,* 109(3), 664-675.

Abdel-Ghaffar, A. M., and Rubin, L. I. (1983b). "Vertical seismic behavior of suspension bridges." *International Journal of Earthquake Engineering and Structural Dynamics,* 11, 1-19.

Abdel-Ghaffar, A. M., and Scanlan, R. H. (1985). "Ambient vibration studies of golden gate bridge: 1. suspended structure, and 2. pier tower structure." *Journal of Engineering Mechanics,* 111(4), 463.

Abdel-Ghaffar, A. M., Scanlan, R. H., and Diehl, J. (1985). *Analysis of the dynamic characteristics of the golden gate bridge by ambient vibration measurements,* 85-SM-1, Department of Civil Engineering, Princeton University, Princeton, NJ.

Analog Devices. (2011). *OBSOLETE PRODUCT — ADXL202,* <http://www.analog.com/en/sensors/inertial-sensors/adxl202/products/product.html> (January 4, 2011).

ANSYS. (2007). *Theory reference for ANSYS and ANSYS workbench,* ANSYS, Inc., Canonsburg, PA.

Buxton-Tetteh, B. (2004). "Development of user cost model for movable bridge openings in Florida." MS Thesis, Florida State University, Tallahassee, FL.

Catbas, F. N., Zaurin, R., Susoy, M., and Gul, M. (2007). *Integrative information system design for Florida department of transportation: A framework for structural health monitoring of movable bridges,* Florida Department of Transportation, Tallahassee, FL.

Catbas, F.N., Zaurin, R., Gul, M. and Gokce, B. (2012a) "Sensor networks, Computer Imaging and Unit Influence Lines for Structural Health Monitoring: A Case Study for Bridge Load Rating," *Journal of Bridge Engineering,* ASCE, 17(4), 662–670, 2012

Catbas, F.N., Gokce, H.B. and Gul, M. (2012b) "Non-parametric Analysis of Structural Health Monitoring Data for Identification and Localization of Changes: Concept, Lab and Real Life Studies," *Structural Health Monitoring Journal,* 11 (5), 613-626

Catbas, F.N., Gul, M. and Gokce, H.B., Frangopol, D. and Grimmelsman, K.A. (2012c) "An Investigation for Issues Related to Condition and Monitoring of Movable Bridges," *Structure and Infrastructure Engineering Journal,* 2012 (accepted)

Chaudhary, M. T. A., Abe, M., Fujino, Y., and Yoshida, J. (2000). "System identification of two base-isolated bridges using seismic records." *ASCE Journal of Structural Engineering,* 126, 1187-1195.

Conte, J. P., He, X., Moaveni, B., Masri, S. F., Caffrey, J. P., Wahbeh, M., Tasbihgoo, F., Whang, D. H., and Elgamal, A. (2008). "Dynamic testing of Alfred Zampa memorial bridge." *Journal of Structural Engineering,* 134(6), 1006-1015.

Darholm, T., Lundh, L., Ronnebrant, R., Karoumi, R., and Blaschko, M. (2007). *Technical book about the Svinesund bridge,* Vägverket (Swedish Road Administration), Göteborg, Sweden.

Goulet, J. A., Kripakaran, P., and Smith, I. F. C. (2010). "Multimodel structural performance monitoring." *Journal of Structural Engineering,* 136(10), 1309-1318.

Grimmelsman, K. A., Pan, Q., and Aktan, A. E. (2007). "Analysis of data quality for ambient vibration testing of the Henry Hudson bridge." *Journal of Intelligent Materials, Systems and Structures,* 18, 765-775.

Gul, M., and Catbas, F. N. (2008). "Ambient vibration data analysis for structural identification and global condition assessment." *ASCE Journal of Engineering Mechanics,* 134(8), 650-662.

He, X., Moaveni, B., Conte, J. P., Elgamal, A., and Masri, S. F. (2009). "System identification of Alfred Zampa memorial bridge using dynamic field test data." *Journal of Structural Engineering,* 135(1), 54-66.

James, G., and Karoumi, R. (2003). *Monitoring of the new Svinesund bridge, report 1: Instrumentation of the arch and preliminary results from the construction phase*, TRITA-BKN Report 74, Brobyggnad, Royal Institute of Technology (KTH), Stockholm, Sweden.

Jonsson, F., and Johnson, D. (2007). "Finite element model updating of the new Svinesund bridge." MS Thesis, Chalmers University of Technology, Göteborg, Sweden.

Karoumi, R., and Andersson, A. (2007). *Load testing of the new Svinesund bridge: Presentation of results and theoretical verification of bridge behavior*, TRITA-BKN Report 96, Brobyggnad, Royal Institute of Technology (KTH), Stockholm, Sweden.

Kim, S., Pakzad, S., Culler, D., Demmel, J., Fenves, G. L., Glaser, S., and Turon, M. (2007). "Health monitoring of civil infrastructures using wireless sensor networks." *6th International Conference on Information Processing in Sensor Networks*, Cambridge, MA.

Koglin, T. L. (2003). *Movable bridge engineering*, John Wiley and Sons, Hoboken, NJ.

Masri, S. F., Miller, R. K., Saud, A. F., and Caughey, T. K. (1987). "Identification of nonlinear vibrating structures: Part I - formulation." *ASME Journal of Applied Mechanics*, 54, 918-922.

Masri, S. F., Sheng, L. H., Caffrey, J., Nigbor, R. L., Wahbeh, M., and Abdel-Ghaffar, A. M. (2004). "Architecture and utilization of a web-based real-time structural health monitoring system." *Smart Materials and Structures*, 13(6), 1269-1283.

Moaveni, B., Barbosa, A. R., Conte, J. P., and Hemez, F. M. (2007). "Uncertainty analysis of modal parameters obtained from three system identification methods." *International Conference on Modal Analysis (IMAC-XXV)*, Orlando, FL.

Nagayama, T., Abe, M., Fujino, Y., and Ikeda, K. (2005). "Structural identification of a nonproportionally damped system and its application to a full-scale suspension bridge." *Journal of Structural Engineering*, 131(10), 1536-1545.

Nayeri, R. D., Masri, S. F., and Chassiakos, A. G. (2007). "Use of eigensystem realization algorithm to track structural changes in a retrofitted building based on ambient vibration." *ASCE Journal of Engineering Mechanics*, 133(12), 1311-1325.

Pakzad, S., Fenves, G., Kim, S., and Culler, D. E. (2008). "Design and implementation of scalable wireless sensor network for structural monitoring." *ASCE Journal of Infrastructure Engineering*, 14(1), 89-101.

Pan, Q. (2007). "System identification of constructed civil engineering structures and uncertainty." PhD Dissertation, Drexel University, Philadelphia, PA.

Plos, M., and Movaffaghi, H. (2004). *Finite element analysis of the new Svinesund bridge: Design model conversion and analysis of the arch launching,* Report 04:12, Chalmers University of Technology, Structural Engineering and Mechanics, Göteborg, Sweden.

Prader, J., Zhang, J., Grimmelsman, K. A., Moon, F. L., Grimmelsman, K. A., and Aktan, A. E. (2010). "Challenges and uncertainty mitigation in structural identification of long span bridges." *The Fifth International Conference on Bridge Maintenance, Safety and Management (IABMAS'10),* Philadelphia, PA.

RTA. (2004). *Design guidelines for bus transit,* Riverside Transit Agency, Riverside, CA.

Schlune, H., Plos, M., and Gylltoft, K. (2009). "Improved bridge evaluation through finite element model updating using static and dynamic measurements." *Engineering Structures,* 31, 1477-1485.

Schlune, H., Plos, M., Gylltoft, K., Jonsson, F., and Johnson, D. (2008). "Finite element model updating of a concrete arch bridge through static and dynamic measurements." *The Fourth International Conference on Bridge Maintenance, Safety and Management (IABMAS' 08),* Seoul, South Korea.

Silicon Designs. (2007). *Low noise analog accelerometer,* <http://www.silicondesigns.com/pdfs/1221.pdf> (January 4, 2011).

Siringoringo, D. M., and Fujino, Y. (2006). "Observed dynamic performance of the yokohama bay bridge from system identification using seismic records." *Journal of Structural Control and Health Monitoring,* 13(1), 226-244.

Siringoringo, D. M., and Fujino, Y. (2007). "Dynamic characteristics of a curved cable-stayed bridge identified from seismic records." *Engineering Structures,* 29(8), 2001-2017.

Siringoringo, D. M., and Fujino, Y. (2008a). "System identification applied to instrumented long-span cable-supported bridges using seismic records." *Earthquake Engineering and Structural Dynamics,* 37, 361-386.

Siringoringo, D. M., and Fujino, Y. (2008b). "System identification of suspension bridge from ambient response measurement." *Engineering Structures,* 30(2), 462-477.

Stahl, F. L., Mohn, D. E., and Currie, M. C. (2007). *The golden gate bridge, report of the chief engineer, volume II,* Golden Gate Bridge and Transportation District, San Francisco, CA.

TinyOS. (2010). *TinyOS home page,* <http://www.tinyos.net/> (January 4, 2011).

Ülker-Kaustell, M., and Karoumi, R. (2006). *Monitoring of the new Svinesund bridge, report 3: The influence of temperature, wind and damages on the dynamic properties of the bridge,* TRITA-BKN. Report 99, Brobygnad, Royal Institute of Technology (KTH), Stockholm, Sweden.

Vincent, G. S. (1958). "Golden gate bridge vibration study." *ASCE Journal of the Structural Division,* 84(ST6), 1.

Vincent, G. S. (1962). "Golden gate bridge vibration studies." *Transactions of the American Society of Civil Engineers,* 127, 667-707.

Wahbeh, M., Tasbihgoo, F., Yun, H., Masri, S. F., Caffrey, J., and Chassiakos, A. G. (2005). "Real-time earthquake monitoring of large scale bridge structures." *International Workshop on Structural Health Monitoring,* Palo Alto, CA.

Wallner, M., and Pircher, M. (2007). "Kinematics of movable bridges." *Journal of Bridge Engineering,* 12(2), 147-153.

The yokohama bay bridge, (1991). Metropolitan Expressway Public Corporation, Tokyo, Japan.

Yun, H., Nayeri, R. D., Tasbihgoo, F., Wahbeh, M., Caffrey, J., Wolfe, R., Nigbor, R. L., Masri, S. F., Abdel-Ghaffar, A. M., and Sheng, L. H. (2007). "Monitoring the collision of a cargo ship with the Vincent Thomas Bridge." *Structural Control and Health Monitoring,* 15(2), 183-206.

Zhang, J., Prader, J., Grimmelsman, K. A., Moon, F. L., and Aktan, A. E. (2009a). "Mitigation of uncertainty in the structural identification of longspan suspension bridges." *The 7th International Workshop on Structural Health Monitoring,* Palo Alto, CA.

Zhang, J., Prader, J., Grimmelsman, K. A., Moon, F. L., and Aktan, A. E. (2009b). "Challenges in experimental vibration analysis for structural identification and corresponding engineering strategies." *International Conference of Experimental Vibration Analysis for Civil Engineering Structures,* Wroclaw, Poland.

Index

Page numbers followed by e, f, and t indicate equations, figures, and tables, respectively.

AASHTO design practice, 72
Accelerometers, 36, 48
Aleatory uncertainty, 3
Alfred Zamba Memorial Bridge (Case Study)
   bridge description and experimental studies, 201, 201f
   data analysis and interpretation, 203–205, 204f, 205f
   experimental studies, 202–203, 202f
   system identification methods used, 202
Ambient vibration survey (AVS), 34
Ambient vibration test (AVT), 29, 30, 33–35
Ammann, Othmar, 174
Analysis errors, 24–25
Anemometers, 43–44, 43f
Anomaly detection, 67, 68
A priori models
   classification of, 18–19
   constraints unique to St-Id, 21–22
   construction using, 22–23, 23f
   development of, 10, 18
   errors in, 24–25
   types of, 19–22, 19f, 20f, 22f
Artificial Neural Networks (ANNS), 6, 71
Autoregressive methods, 69–70
Autoregressive models (AR), 67, 85–86
Autoregressive moving average model with exogenous input (ARMAX), 86
Autoregressive moving average vector (ARMAV) models, 6

Base-excited structures, 85–86
Bridges. *See* Structural identification of bridges case studies

Building Substructure Example: Composite Structural Floor system (Case Study)
   data analysis and model calibration, 159–161, 160f, 161f
   experimental studies, 157–158, 157f–159f
   interpretation and decision making, 161–163, 162f, 163f
   model development, 156–157, 156f, 157f
   objective of St-Id application, 155
   structure description, 155, 156f

Calibration. *See also* Model calibration techniques
   finite element model, 87–95
   function of, 78–79
California Integrated Seismic Network (CISN), 135
California Strong Motion Instrumentation Program (CSMIP), 119, 130, 131, 133, 133f, 134
Case studies. *See* Structural identification of bridges case studies; Structural identification of buildings case studies; *specific case studies*
CCD arrays, 38
Chicago Full-Scale Monitoring Program (Case Study)
   data analysis and model calibration, 124–125, 124t, 125f
   experimental studies, 123, 123f
   interpretation, 125–126, 126f
   model development, 121–123
   objectives of ST-ID application, 121
   program description, 119, 121

Complex Exponential Algorithm (CEA), 82–83
Constructed systems. *See also* Structural identification of constructed systems (St-Id)
  experimentation on, 10–11
  overview of, 1–3
  performance-based engineering for, 114–115, 114t
  simulating performance of, 1
  uncertainties unique to, 2, 3t
Controllable dynamic input, 28, 29–33
Controllable static input, 28–29
Controlled load tests, 30
Controlled traffic, 30
Correlation analysis, 67
Corrosion, monitoring of, 44–45

DAKOTA (Sandia National Laboratories), 7
Damage Locating Vector (DLV) technique, 82
Data acquisition
  explanation of, 48
  for modal surveys, 49–50
  for short/long term structural monitoring, 48–49
Data interpretation
  benefits of, 65
  challenges related to, 65–66
  non-physical numerical models and, 67–72
  types of, 66
Data mining, 70–71
Data processing, 11
Data reduction, 68
Data storage, 50
Data transmission, 47–48
Decision-making process
  risk-based, 115–116, 116f, 117t
  for St-Id applications, 113
Deck profiling systems, 41
Direct data interpretation, function of, 11
Direct stress measurement, 42
Discrete-time filter design, 85
Displacements
  from acceleration, velocity, strain or rotation signals, 41, 42f
  studies involving, 36–37
Distributed Brillouin Scattering Sensors, 47
Distributed Raman Scattering Sensors, 47
Drop-weight systems, 32, 34f
Dynamic input
  controllable, 28, 29–33
  uncontrollable, 28, 33
  uncontrollable unmeasurable, 28, 34–35

Eigenstructure assignment method, 94
Eigensystem Realization Algorithm (ERA), 83
Electro-dynamic shakers, 31, 31f
Electro-mechanical impedance (EMI) technique, 44
Electronic distance measurement (EDM), 39
Empirical Mode Decomposition (EMD), 6
Evolutionary Algorithms (EAs), 94, 95
Evolutionary optimization, 94–95
Experimental modal analysis (EMA), 34, 159–161
Experiments/experimentation
  background of, 26–27
  classification based on input, 28–35
  components of, 27–28
  controlled, 10–11
  data acquisition and management, 48–50
  data transmission, 47–48
  non-destructive evaluation applications, 51–52, 51t
  sensors and sensor classification based on measurand, 35–47
Extrinsic Fabry-Pérot Interferometer (EFPI), 46

Fabry-Pérot Interferometric sensors, 46
Feature selection techniques, 71–72
FE codes, 21

STRUCTURAL IDENTIFICATION OF CONSTRUCTED SYSTEMS   229

FE software, 21
Fiber Bragg Grating (FBG) arrays, 35, 36
Fiber Bragg Grating (FBG) sensors, 45
Fiber optic connections, 48
Fiber optic sensors (FOS)
  function of, 45
  types of, 45–47, 46f
Finite element analysis, 23–24
Finite element model calibration
  background of, 87–88
  error characterization and, 88–89, 89t
  objective functions for error minimization and, 89–91
  techniques of, 91–95
Finite element modeling
  in Chicago Full-Scale Monitoring Program case study, 121–122
  explanation of, 21, 22f
  in Seven-Story RC Building Slice case study, 153–154
Forced vibration tests (FVTs), 29, 30, 32, 34
Fourier Transform, 68, 84
Four Seasons Building (Case Study)
  building description, 126, 127f
  data analysis and model calibration, 128–129, 128t, 129f
  experimental studies, 127, 128f
  interpretation and decision making, 129f, 130
  model development, 127
  objective of St-Id application, 126–127
Frequency Domain Decomposition (FDD), 85
Frequency response function (FRF) data, 30, 32
Frequency response function (FRF) matrix, 32
FRF assignment, 94

Global Positioning System (GPS), for tracking behavior of structures, 39–41

Golden Gate Bridge (Case Study)
  bridge description, 181, 181f
  comparison and interpretation, 186–188, 187f
  experimental program, 182–184, 183f, 184f
  objectives of St-Id application, 182
  system identification and data analysis, 184–186, 185f, 186f
Guangzhou New TV Tower (Case Study)
  data analysis and model calibration, 145–146, 145f, 146f, 146t, 147t
  model development, 141f, 142–144, 143f–144f
  objectives of St-Id application, 140–141
  tower description, 140

Hakucho Suspension Bridge (Case Study)
  bridge description, 193
  data analysis and system identification, 195
  instrumentation and ambient vibration monitoring, 194, 194f
  objective of St-Id application, 193–194, 194f
  structural identification and interpretation, 195–196, 196f
Hammer testing, 32
Heisenberg Uncertainty Principle, 68
Henry Hudson Bridge (Case Study)
  bridge description, 169, 169f
  data analysis and model calibration, 171–173, 172t, 173f
  experimental studies, 170–171, 171f
  objective of St-Id application, 169
  system identification method development, 169–170, 170f
High ingress protection (IP), 47
Hilberg-Huang Transformation, 6
Hybrid carbon fiber-reinforced polymer (HCFRP) sensing techniques, 47

Hybrid matrix update methods, 94

Ibrahim Time Domain (ITD) method, 83
Image tracking applications, 37–38
Impact hammers, 32
Impact testing, 32
Impulsive testing, 32
Incontrollable static input, 28, 29
Input
  dynamic, 28–35
  static, 28–29
Input errors, 24
In-service structural performance, 26
Instance-Based Method (IBM), 67
Instrumental hammers, 32
Instrumented buildings inventory, 119, 120t

Langensand Bridge (Case Study)
  bridge description, 205, 206f
  data analysis and structural identification, 208, 208f
  experimental studies, 207, 208f
  model development, 207, 207f
  objectives and structural identification, 206
Laser-based displacement technologies, 37
Laser Doppler Vibrometers (LDVM), 37
Laser interferometry, 37
Least-Square Complex exponential algorithm (LSCEA), 83
Level monitoring systems, 41
Linear mass exciters, 31–32, 31f
Load testing, 4
Low smoke zero halogen (LSZH), 47

Maillart, Robert, 4
Mechanical impedance, 44
Microwave interferometry, 40, 40f
Modal analysis
  data acquisition for, 49–50
  experimental, 34
  operational, 34
Modal coordinates, 80

Modal models
  function of, 79–80
  identification using, 82–83
Model calibration techniques. *See also* Calibration
  in Chicago Full-Scale Monitoring Program case study, 124–125
  eigenstructure assignment, 94
  evolutionary optimization, 94–95
  explanation of, 91
  hybrid matrix update, 94
  manual, heuristic-based, 91–92
  optimal matrix update, 92–93
  sensitivity-based update, 93–94
Modeling errors, 24, 88–89
Models
  modal, 79–80
  use of multiple, 79
  use of single, 78–79
Model selection
  explanation of, 78
  modal models and, 79–80
  structural models and, 78–79
Morlet wavelet, 69
Multiple input multiple output (MIMO), 32, 83, 85

NESSUS (Southwest Research Corporation), 7
New Svenesund Bridge (Case Study)
  bridge description, 217, 217f
  data analysis and model calibration, 219–222, 220f, 221f
  experimental studies, 219
  model development, 218, 218f
  objective of St-Id application, 218
Newton-Raphson method, 93
Nondestructive evaluation (NDE)
  advances in, 10
  for structural identification, 51–52, 51t
Non-physical numerical models
  anomaly detection and, 67
  data reduction and representation and, 68–71

STRUCTURAL IDENTIFICATION OF CONSTRUCTED SYSTEMS 231

examples of, 67
feature selection and extraction and, 71–72
Non-physics-based models (NPMs)
 data interpretation and, 66
 nature of, 18–19
 research and development based on, 5, 5t, 6
 utilization of, 12

Operational modal analysis (OMA), 34
Optical marker tracking, 38
Optimal matrix update methods, 92–93
Optometer, 37–38, 38f

Parameter identification
 approaches of, 80–82, 81t
 modal, 91
Parameter sensitivity based model updating, 91–92
Pedestrians, 35
Performance-based engineering, 114–115, 114t, 119
Performance limit states, 116, 117t
Perturbation matrix, 93
Phenomenological models, 19–20, 19f
Photogrammetry, 9
Physics-based models (PBMs). See also A priori models
 data interpretation and, 66
 nature of, 18, 19
 research and development based on, 5, 5t, 6
 selection and calibration of, 11–12
 utilization of, 12
Plato, 4
Pneumatic systems, 41
Polyreference Complex Exponential technique, 83
Polyreference Frequency Domain (PFD), 83
Polyreference Time Domain technique, 83
Power spectral approaches, 84–85
Pressure measurement technology, 43

Principal Components Analysis (PCA), 70–71
Proportional Flexibility Matrix (PFM), 82
Pulse-train generation, 33
PZT patches, 44

Qualitative risk assessment, 118
Quality control requirements, 23–25
Quantitative risk assessment, 117–118

Random Decrement Signature (RDS), 84
Random Decrement Technique (RDT), 83, 84
Rational polynomial models, 67
Real time kinematic (RTK), 39
Reciprocating mass exciters, 31–32, 31f
Research and development, on structural identification, 5–7, 5t
Risk
 actual and perceived, 116, 116f
 definition of, 115
Risk assessment, quantitative vs. qualitative, 117–118
Risk-based decision-making, 115–116, 116f, 117t
RMS errors, 90
Rotating eccentric mass (REM) exciters, 30–31

Sandia National Laboratories, 7
Scalogram, 69
Seismic excitation, 33
Sensitivity-based update methods, 93–94
Sensors, 35–36
Seven-Story RC Building Slice (Case Study)
 building description, 148, 149f
 damage identification through finite element model updating, 153–154, 154f, 155f
 data analysis, 150–152, 151f–153f
 experimental studies, 150
 objective of St-Id application, 148

system identification methods applied, 149
Shaker testing, 32
Short-Time Fourier Transform (STFT), 68
Single input multiple output (SIMO), 83
SOFO interferometric sensors, 45–46, 46f
Southwest Research Corporation, 7
Spectral analysis, 68
Spectrogram, 68
State space models, 6
Static input, 28–29
Statistical significance tests, 67
Step relaxation, 33
St-Id. See Structural identification of constructed systems (St-Id)
Stiffness matrix, 82
Stochastic Damage Locating Vector (SDLV), 82
Strain gauges, 42
Structural identification of bridges case studies
   Alfred Zampa Memorial Bridge, 201–205, 201f–205f
   Golden Gate Bridge, 181–188, 181f, 183f–187f
   Hakucho Suspension Bridge, 193–196, 194f, 196f
   Henry Hudson Bridge, 169–173, 169f–171f, 172t, 173f
   Langensand Bridge, 205–208, 206f–208f
   New Svinesund Bridge, 217–222, 217f, 218f, 220f, 221f
   Sunrise Boulevard Movable Bridge, 209–217, 209f–217f, 213t, 216t
   Throgs Neck Bridge, 174–179, 174f–178f, 179t, 180f
   Vincent thomas Bridge, 188–193, 189f–193f
   Yokohama Bay Bridge, 196–200, 197f, 198f, 200f
Structural identification of buildings case studies
   background, 119, 120t
   Building Substructure Example: Composite Structural Floor System, 155–163, 156f–163f
   Chicago Full-Scale Monitoring Program, 119, 121–126, 123f, 124t, 125f, 126f
   Four Seasons Building, 126–130, 127f–129f, 128t
   Guangzhou New TV Tower, 140–148, 141f, 143f–146f, 146t, 147t
   Seven-Story RC Building Slice, 148–154, 149f, 151f–155f
   Three-Story concrete Building in CSMIP, 130–140, 130f, 132t–134t, 133f–137f, 138t, 139f
Structural identification of constructed systems (St-Id)
   applications for, 2–3, 9
   for assessment and decision making, 113–118
   challenges related to, 95–96
   experimental considerations and, 10–11, 26–28 (See also Experiments/experimentation)
   explanation of, 1–2, 5
   finite element model calibration and, 87–95, 87f, 89t
   function of, 5
   historical background of, 4–7, 5t
   model application for, 80–86, 81t
   modeling constraints unique to, 21–22
   model selection for, 78–80
   nondestructive evaluation for, 51–52, 51t
   overview of, 1–3, 3t
   physics- and non-physics-based approaches to, 5, 5t, 6
   potential of, 2
   research and development and, 5–7, 5t

STRUCTURAL IDENTIFICATION OF CONSTRUCTED SYSTEMS   233

Structural identification of constructed systems (St-Id) steps
background of, 7–8, 7f, 12–13
data processing and feature extraction as, 11
experimentation control as, 10–11
measurement, visualization and a-priori modeling as, 9–12
objectives, observation, and conceptualization as, 8–9
selection and calibration of physics-based models as, 11–12
utilization of models for decision making as, 12
Structural models
explanation of, 20, 20f, 78
identification using, 81–82
Subspace State-Space System Identification (N4SID), 128
Sunrise Boulevard Movable Bridge (Case Study)
bridge description, 209, 209f
data analysis and model calibration, 211–214, 212f–214f
experimental studies, 210, 210f
interpretation and decision making, 214–217, 215f, 216f
objective of St-Id application, 209–210
system identification method development, 211, 211f
Support vector machines (SVM), 71–72
Surveying techniques, for tracking structures, 39–40
Swinging bell tests, 33
Swinging train tests, 33

Temperature sensors, 43
3D CAD, 22–24, 23f
3D visualization, 9
Three-Story Concrete Building in CSMIP (Case Study)
data processing and archiving, 134–135, 135f–137f
instrumentation overview, 133, 133f, 134f

interpretation, 137–140, 138t, 139f
objective of St-Id application, 131
program description, 130–131, 130f
system identification method, 131–132, 132t, 133t
Throgs Neck Bridge (Case Study)
bridge description, 174, 174f
data analysis and model calibration, 179, 179t, 180f
experimental studies, 177–179, 177f, 178f
objective of St-Id application, 174
system identification method development, 174–276, 276f
Tracking lasers, 37
Traffic excitation, 34–35
Transient testing, 32

Uncertainty models, 89t
Uncontrollable dynamic input, 28, 33
Uncontrollable measurable dynamic loads, 33
Uncontrollable unmeasurable dynamic input, 28, 34–35

Velocity measurements, 41–42
VHS lasers, 37
Vincent Thomas Bridge (Case Study)
bridge description, 188–189, 189f
bridge identification using seismic vibration data, 191–192, 191f
forensic study of ship-bridge accident using St-Id, 192–193, 192f, 193f
objective of St-Id application, 190–191

Wave excitation, 35
Wavelet decomposition, 6
Wavelet Transform (WT), 68–69
Wind excitation, 34
Wind measurement, 43–44, 44f
Wind tunnel models, in Chicago Full-Scale Monitoring Program case study, 122–123

Wired technology, 47–48
Wireless monitoring systems, 48

Yokohama Bay Bridge
  bridge description, 196–197, 197*f*
  data analysis and interpretation, 199–200, 200*f*
  experimental studies, 198, 198*f*
  objective of St-Id application, 197
  system identification method development, 197–198

Z24 tests, 32